COMBAT LEGEND

B-17

FLYING FORTRESS

Martin W. Bowman

Airlife

Text written by Martin W. Bowman
Profile illustrations by Dave Windle
Cover painting by Jim Brown – The Art of Aviation Co. Ltd

First published in the UK in 2002
by Airlife Publishing Ltd

British Library Cataloguing-in-Publication Data
    A catalogue record for this book
    is available from the British Library

ISBN 1 84037 365 2

Printed in Hong Kong

For a complete list of all Airlife titles please contact:

Airlife Publishing Ltd
101 Longden Road, Shrewsbury, SY3 9EB, England
E-mail: sales@airlifebooks.com
Website: www.airlifebooks.com

Contents

Fortress Timeline

18 July 1934: AAC issues specification for a bomber to replace the B-10
17 July 1935: Model 299 rolls out at Boeing Field, Seattle
30 October 1935: Model 299 crashes
2 December 1936: First flight of Y1B-17, 36-149
March-September 1937: first Y1B-17s received by 2nd Bomb Group
15 October 1937: First flight of the XB-15
15-27 February 1938: Fifteen Y1B-17s of 2nd BG make goodwill trips to South America
12 May 1938: Three Y1B-17s intercept the Italian liner *Rex* 725 miles east of New York City
27 June 1939: First B-17B flies
30 August 1940: B-17E ordered into production
8 July 1941: Three RAF Fortress Is attack Wilhelmshaven
21 July 1940: First B-17C flies
October 1939-March 1940: First B-17Bs delivered to the 2nd and 7th Bomb Groups

1941
5 September: First B-17E flies
11 September: President Roosevelt endorses Air War Plan (AWPD/1)
27 October: B-17B of 2nd BG attacks a U-boat
7 December: Many Fortresses lost at airfields in Hawaii
8 December: USA declares war on Japan. Japanese air strikes on Clark and Iba airfields, Philippine Islands. Many Fortresses among the losses
11 December: Germany declares war on the USA

1942
23 February: 8th AF arrives in England
2 April: Andaman Islands attacked by B-17s
23 April: Brigadier-General Ira C. Eaker commands VIII Bomber Command in Europe
7-8 May: Battle of the Coral Sea. Japanese threat to New Guinea thwarted
30 May: First B-17F delivered to the AAC
4-6 June: Battle of Midway reverses Japanese tide in the Pacific. 7th AF B-17s and B-26s help in repelling invasion fleet
11 June: 11th AF bombers make first attack on Kiska, main Japanese air base in the Aleutians
8 August: Pacific Offensive begins with landings on Guadalcanal, Solomon Islands. AAF B-17s attack Rabaul to divert enemy
17 August: First American heavy bomber raid against railyards at Rouen, France, by B-17s of the 97th BG
August-September: Allied Air Forces, SWPA, renamed 5th AF, help repulse Japanese thrusts at Buna in New Guinea
October: US aircraft in the Pacific start sowing mines in narrow waters, river mouths and harbours all over Japanese-held territory
20 October: VIII Bomber Command targetted against U-Boat bases
8 November: US 12th AF supports *Torch* landings in N Africa
December: Southeastern New Guinea secured

1943
23 January: Casablanca Conference marks the start of the Combined Bomber offensive in Europe; USAAF by day, RAF Bomber Command by night
27 January: First US attack on Germany (U-boat construction yards at Wilhelmshaven).Only three bombers shot down
13 February: Last B-17 misions in Aleutians before redeployment to South Pacific. By mid-1943 most B-17s withdrawn from the Pacific
1-2 May: Battle of the Bismarck Sea: 3,000 Japanese troops lost as twelve enemy ships are sunk by 5th AF planes
13 May: Axis surrender in North Africa
21 May: First B-17G flies
June: Operation *Pointblank*, against German submarine yards and bases, aircraft industry, ball bearings and oil, is begun
22 June: First deep penetration of Germany; 6 out of 235 B-17s lost and 170 damaged in successful attack on synthetic rubber plant at Huls
10 July: Allied invasion of Sicily
17 July: 322 8th AF bombers despatched to Hanover
24 July: 'Blitz Week' begins and costs 8th AF 100 aircraft and 90 combat crews – fewer than 200 heavy bombers ready for combat
15 August: *Starkey* deception plan created to make Germany believe that an invasion of the French coast is imminent
17 August: 376 8th AF B-17s bomb the Schweinfurt ball bearing plant and the aircraft plants at Regensburg. Heavy losses
September: 5th AF assists Army forces in capture of Lae, Finschhafen, in New Guinea
9 September: Allied invasion of Italy. Italy surrenders. Precision bombing of Fw 190 plant at Marienberg by 96 B-17s of the 8th AF
1 October: 15th AF begins strategic bombing of German and Austrian targets from Italy
14 October: Second Schweinfurt raid: 291 B-17s attack, 60 aircraft lost, and 138 damaged
2 November: Devastating US raid on Japanese base at Rabaul

1944
1 January: US Strategic Air Forces in Europe Command established to control operations of 8th and 15th AFs
19-26 February: 'Big Week' series of sustained raids on German aircraft industry. Total losses amount to 226 US bombers
4-6 March: B-17s of the 8th AF become the first US bombers to attack the German capital
March-June: Interdiction campaign to isolate northwestern France, the area of Operation *Overlord*
19 March-11 May: Operation *Strangle*: interdiction campaign to choke off German rail supply in Italy
2 June: First *Frantic* shuttle mission to Russia. Lt-Gen Eaker leads a 15th AF shuttle force of 130 B-17s and 70 P-51s from Italy
6 June: Operation *Overlord*, the invasion of Normandy: 2,362 bomber sorties involving 1,729 B-17s and B-24s are flown on D-Day
11 June: Return shuttle raid between the USSR and Italy. 15th AF bombs the Romanian airfield at Foscania
21 June: Second *Frantic* shuttle mission, by 8th AF B-17s, escorted by P-51s. Bombers stage through Italy; return to England on 5 July
14 July: Operation *Cadillac*: mass drop of supplies to French Forces of the Interior by 322 8th AF B-17s
6 August: 76 Fortresses and 64 Mustangs fly shuttle mission to Russia, bombing the Focke Wulf plant at Rahmel, Poland, en route. Return via Italy, 8 August, returning to Britain on 12 August
11 September: Final *Frantic* shuttle mission begins
18 September: 117 B-17s of the 8th AF drop ammunition, guns and supplies to Poles in Warsaw, but most are taken by Germans

1945
January: US tactical air forces help defeat German forces engaged in the 'Battle of the Bulge', the last enemy offensive in the west
February: Raids on Berlin: bombing offensive continues through to the last days of the war
13-15 February: 8th AF and RAF Bomber Command raze city of Dresden. HE and firestorms cause 35,000 civilian deaths
7 May: German representatives surrender in the west. War ends. VE Day proclaimed on 8 May 1945. Russians accept surrender, 9 May

1. America's Flying Fortress: Development History of the B-17

In the 1930s it was an accepted theory among aviation strategists that a formation of unescorted bombers could get through to their target if they were properly arranged and adequately armed. During American air manoeuvres in 1933, pursuit fighters repeatedly failed to intercept the bombers, and there was even talk of eliminating pursuits altogether. Funds for new aircraft were very limited and it was mostly manufacturers who funded new developments, which in turn might attract orders from the military.

In 1934 Congress passed legislation which forced aircraft and engine manufacturers to sever the links they had forged with airline operations. The Boeing Aircraft Company resumed independent operation and moved into the bomber business. On 14 April 1934, Boeing and Martin received the US Army's General Staff request for design proposals for a research model. It was part of 'Project A', which was intended to produce an aircraft capable of flying non-stop to Hawaii, Panama and Alaska. This aircraft, the largest military type proposed up to that time, would have a wingspan of about 150 feet and weigh some 30 tons.

'Project A' was so massive mainly because it needed storage for the enormous fuel load to carry a 2,000-lb bomb load 5,000 miles at 200 mph. Boeing proposed the Model 294 or the XBLR-1 (Experimental Bomber, Long-Range), as it was first known. It was designed to be powered by four 1,000 hp Allison V-1710-3 liquid-cooled in-line engines, and was armed with three .30-calibre and three heavy .50-calibre machine-guns. These were housed in a nose turret, a forward-facing belly turret below the pilot's cabin, a top turret, two waist blisters behind the wings, and a rearward-facing belly turret.

On 28 June 1934 Boeing was awarded the contract for design data, wind tunnel tests and a mock-up. This was followed, on 29 June 1935 by a contract to build the XB-15 flying example, as it was known. Boeing's only other rival, the 52 ton Martin Model 145/XB-16, which was to have a span of 173 feet and be powered by six V-1710-3s, four as tractors, two as pushers, was too expensive to build.

Massive long-range bomber

The XB-15 had a 149 foot span, and weighed some 70,700 lb. It had passageways in the wing leading edges giving access to four Pratt & Whitney R-1830-11 air-cooled Wasp radials – the Allison engines for which the aircraft had been designed were as yet unavailable. It had double wheels and two ailerons on each wing. A crew of ten had soundproofed, heated and ventilated quarters with rest bunks, kitchen and toilet, and, for the first time in an aircraft, small auxiliary engines produced a 110-volt electrical system.

The XB-15 flew for the first time on 15 October 1937. As completed, it could carry a 12,000-lb maximum bomb load, or 2,511 lb over 3,400 miles. Although greatly underpowered for its size, Boeing anticipated improving performance using 1,400-hp engines on the YB-20 production model. However, these two improved test versions were later cancelled. The XB-15 joined the 2nd Bomb Group in August 1938 and in 1942 was converted to the XC-105 cargo carrier. Known affectionately to its operators as 'Grandpappy', it was used to haul cargo until 1945.

While Boeing wrestled with the problems associated with the experimental monster, on 18 July 1934 the US Army Air Corps (Air Corps) issued a specification for the next production four- to six-place bomber to replace the then-standard twin-engined Martin B-10. On 8 August 1934 Boeing were invited to submit a company-owned design in a fly-off bomber competition to be held in 1935.

At the time the invitation was received, Boeing designers were working towards two new designs, the Model 299 bomber and the Model 300 civil transport project. These two designs would have many common parts but at first the Model 300 was given priority since a new airliner which could challenge the Douglas

DC-2 was needed to replace the Boeing 247. However, once Boeing received confirmation that the new 'multi-engined' bomber could have four engines instead of the more usual two, the company decided to abandon the Model 300 for the time being, and concentrate all its design effort on the Model 299. In the event, the Model 300 was never built, but the Model 307 33-passenger transport, which used the B-17 wings, tail, powerplants, and landing gear mated to a pressurized fuselage appeared in December 1938.

On 16 September 1934 Boeing boldly decided to invest $275,000 of its company capital in the Model 299, despite cash on hand at this time standing at less than $500,000. It would be built in the 1916 vintage factory on the Duwamish River. As manufacturers they would have to build prototypes at their own expense, and a flying prototype – capable of carrying a 2,000-lb bomb load at a speed of 200–250 mph over a distance of 1,020–2,000 miles – had to be

X13372, the Model 299, at its roll-out at Boeing Field, Seattle, 17 July 1935. Because its wingspan was greater than the width of the hangar door, it had to be rolled out sideways on wheeled dollies. The Model 299 was flown for the first time on 28 July by the company test pilot Leslie Tower. The clean lines of the new bomber owed much to the sleek Model 247 airliner which had been scaled up into the much bigger Model 299. Many of the engineering innovations employed had been developed on the earlier Model 294 (XB-15) project. (Boeing)

X13372, the original Model 299 prototype, gets airborne for the first time on 28 July 1935. Built at the Boeing factory on the Duwamish River, it was trucked to Boeing Field, Seattle where it was assembled. At the controls for the first flight was test pilot Leslie Tower. *(Boeing)*

available for trials at Wright Field in August 1935. The winner could expect to receive an order for some 220 bombers.

The Model 299 design owed much to the company's earlier models, the B-9 monoplane bomber of 1931 and the Boeing 247 transport of 1933. Further data was available as a result of the work on the XB-15. The project engineer was E. G. Emery, with Edward C. Wells as his assistant. In December 1935 Wells would be promoted to succeed Emery.

Extremely strong

Tubular strutting was used in the construction of the Model 299 (X13372) to produce a structure of great strength. The fuselage was a conventional semi-monocoque all-metal structure of basically circular configuration, consisting of four main sub-assemblies. These in turn were made up of nine sub-assemblies riveted together into stressed elements. A series of circumferential frames and vertical bulkheads with longitudinal stringers and covered with stressed skin provided a very strong structural unit.

Eighteen sub-assemblies made up an extremely efficient wing with a low weight/strength ratio. Truss-type main spars

were capped with sheet metal and gusseted girders. Sections between spars were covered with corrugated aluminium sheet, and stressed skin was riveted to corrugated areas and to the tube and channel truss-type ribs. The entire structure produced a wing with an exceptional ability to absorb punishment without loss of structural integrity. A symmetrical NACA aerofoil was used. Ailerons were of all-metal structure with fabric (and later aluminium skin) covering while the tail surfaces were made up of cantilevered stabilizers. Sheet metal covered the stabilizers, while fabric was used on the rudder and elevators.

The Model 299 was the first US aircraft to have air brakes in the landing-gear wells. The unique landing gear was an electrically operated, twin yoke arrangement with a retracting screw. Each main gear had to be operated separately. The tail wheel too was retractable, and all three wheels, when retracted, were half exposed. The Model 299 was also the first to be equipped with flaps on the rear edges of its wings to serve as air brakes on landing. The split flaps were of all-metal construction. The landing lights, which were in protruding cylinders, were later faired into the wings on subsequent models.

When the XB-15 35-277 made its first flight on 15 October 1937, twenty-eight months after the Model 299, it was the largest aircraft in the world. The new bomber, which was armed with six machine-guns, also contained complete living and sleeping quarters with soundproofed, heated, ventilated cabins. The XB-15 joined the 2nd Bomb Group in August 1938 and established a number of records. In 1942 the aircraft was made into a transport with cargo doors and hoist. Redesignated the XC-105, it was popularly known as 'Grandpappy'. The XC-105 continued to serve until 1945, when it was dismantled. *(Boeing)*

The Model 299, unlike the B-9 bomber, was designed to carry all bombs internally. The bomb load amounted to 4,800 lb in eight vertical stacks of 600 lb bombs each, and 2,500 lb of bombs could be carried a distance of 2,040 miles. Four streamlined machine-gun cupolas bulged from the sides, top and bottom of the sleek fuselage, which tapered to a shark-fin vertical tail. The cupolas, together with the nose gunner's station, all contained single-mount .30- or .50-calibre machine-guns for defence.

Power was provided by four 750-hp Pratt & Whitney S1E-G (R-1690) 'Hornet' commercial radials, housed in NACA cowlings faired into the nacelles ahead of the wing.

Completed in under a year, the Model 299 was barged across the Duwamish and towed without wings through the city streets of Seattle to Boeing Field, the King County Airport, on 17 July 1935. After assembly and brief taxiing tests, the Model 299 was flown for the first time on 28 July 1935 by the company test pilot, Leslie

Testing was almost complete and the Air Corps was about to confer the title XB-17 to the Model 299 when, on 30 October 1935, the aircraft crashed with Major Ployer P. 'Pete' Hill (chief of Wright Field's Flight Testing Section) at the controls. Hill died later in the day and Leslie Tower, Boeing test pilot, died a few days later. The subsequent investigation concluded that the crash was a result of the mechanical ground locks not having been disengaged prior to take-off. This meant that the aircraft's control surfaces were frozen on take off. *(Boeing)*

Tower. Richard L. Williams of the *Seattle Times*, on being told that the machine was to be used in a defensive role to protect the American coastline from enemy surface fleets, dubbed the fifteen-ton bomber, 'Flying Fortress'.

At first, testing went very well indeed. On 20 August 1935, Tower, his assistant and co-pilot Louis Wait, with C. W. Benton as mechanic and Henry Igo of Pratt & Whitney on board to maintain the engines, flew the Model 299 from Seattle to Wright Field at Dayton, Ohio, to begin service trials in competition with the twin-engined Martin 146 and the Douglas DB-1. The Model 299 completed the 2,100-mile trip, much of the way on auto-pilot, in a record breaking nine hours non-stop with an unbelievable average speed of 233 mph.

Air Corps pilot Lt Donald Putt was assigned to the Model 299 as project test pilot. Competitive testing soon proved that the Boeing aircraft was far superior to both the twin-engined entries, the Martin and the DB-1, which basically were an improved B-10, and a DC-2 civil transport with a bomber fuselage respectively. The Boeing entry went on to exceed all the Army specifications for speed, climb and range. Major Ployer 'Pete' Hill (chief of Wright Field's Flight Testing Section) took over the final tests from Putt.

Testing was almost complete when on 30 October 1939 Pete Hill took the controls for yet another flight. Putt sat beside him in the right-hand seat and Leslie Tower stood behind them on the flight deck. The Model 299 raced down the runway and began its climb before appearing to stall. It crashed and burst into flames. Putt, Benton and Igo scrambled clear of the wreckage but Major Hill and Tower were trapped inside. They were bravely pulled clear by other members of the crew, but Hill never regained consciousness and died later in the day. Tower died a few days later.

Pilot error
An investigation concluded that the crash was a result of the mechanical ground locks not having been unlocked prior to take off. This prevented movement of the main surfaces. No

The six Y1B-17s of the 2nd Bomb Group, which made the Goodwill flight from Langley Field, Virginia, to Buenos Aires, in February 1938, pictured over the New York City skyline. They covered a total of 12,000 miles without serious incident. At a time when most bombers were twin-engined, the B-17 made an excellent and very impressive 'Show the Flag' ambassador for the United States. *(Boeing)*

blame could possibly be attributed to Boeing; obviously it was pilot error. However, the outcome was that on 17 January 1936 Boeing received only a service test order for thirteen improved B-299Bs and a static test model under the designation YB-17. Before the crash the Air Corps had been considering an order for 65 B-17 bombers. Douglas on the other hand, received an order for 133 B-18 Bolos, followed, in June 1937 by an order worth $12 million for 177 improved B-18As, plus a further 40 B-18As in mid-1938.

More power

The YB-17, which changed to Y1B-17 on 20 November 1936 because of budget transfers from Army Air Corps (AAC) appropriated funds to Category F-1 funds, was ready for test flying on 2 December 1936. The most significant change comparedto the Model 299 was the introduction of 850-hp (1,000-hp on take-off)

military Wright SGR-1820-39 'Cyclone' engines in place of the earlier Hornets. The landing gear was also changed to a single leg arrangement instead of the earlier double legs, and minor changes were made to the armament systems. The crew was increased to nine.

Boeing had built a new factory on Boeing Field for B-17 production but the first few Y1B-17s were built at the old Plant 1 and barged across the river to the airport two miles away. On 2 December 1936 Major John D. Corkville, Capt Stanley Umstead and a five-man crew flew Y1B-17 36-149, the first Y1B-17, from Boeing Field. The 55 minute test flight went without a hitch, as did the second test hop two days later. On the third test flight, on 7 December, Umstead was at the controls of 36-149. The flight almost ended in complete disaster. While taxiing, Umstead applied the brakes so hard that the early type disk brakes overheated. He compounded his error after

Y1B-17A 37-369, seen here on Boeing Field at Seattle on 30 January 1938, was originally intended as a static test aircraft. However, it was subsequently redesignated Y1B-17A and used to test the supercharged R-1820-51 engines which would give the B-17 its outstanding high-altitude performance. *(Boeing)*

getting airborne. Instead of leaving the gears down to let the brakes cool he retracted the wheels immediately so that in a very short time the red-hot, half molten brake plates had welded themselves into a solid mass.

A sorry sequence of events then began to overtake Umstead. During the short flight one of the Cyclone engines began to overheat and it had to be shut down. On the return leg to Boeing Field a second engine also overheated and had to be shut down. Umstead, unaware of the problems he had caused to the brakes, touched down and was shocked to discover that the wheels had locked solid. The stuck wheels brought the Y1B-17 to an abrupt halt and stood the aircraft on its nose on the runway. Luckily, no one was hurt. Equally fortunate for Boeing, the subsequent Congressional investigation resulted in no action being taken against the company. The aircraft was repaired, and was flying again by 2 January 1937. Rubber de-icer boots were fitted to the leading edge of the wing, and aluminium covering was substituted for fabric on the flaps.

Ironically, the first Y1-17B was delivered to the AAC more than two months before the first Douglas B-18 Bolo. By 5 August 1937, the last of the 13 Y1B-17s had been delivered to the 2nd Bombardment Group at Langley Field, Virginia, where they went into service with the 20th, 49th and 96th Bombardment Squadrons. Lt-Col Robert C. Olds and his officers and men pioneered the early use of the B-17 and it gained a well-earned reputation for rugged construction and safe operation. There were no serious mishaps during three years of intense military flying.

During this period many improvements were made, not least the introduction of pre-flight checklists to prevent a repetition of the Model 299 crash. Many officers who would soon become famous in World War Two, such as Harold L. George, Curtis E. LeMay, Robert B. Williams, Neil Harding and Caleb Haynes, served in the 2nd Bomb Group.

Showing the flag

On 15 February 1938, six Y1B-17s led by Lt-Col Olds made a very successful goodwill trip to South America, visiting Peru and Buenos Aires, Argentina, with the return on 27 February. Later, another seven Y1B-17s, led by Major-

Seen here at Seattle on 27 July 1939, this is one of the first thirty-nine improved B-17Bs built. Except for minor changes in the fairing of the machine-gun blister into the fuselage, the remainder of the B-17B armament was the same as used on the Y1B-17. The offset aircraft commander's blister behind the cockpit was later moved to the centre line on the B-17D. Internally, some crewmembers were relocated and improved R-1850-51 engines delivered 1,200 hp for take-off. *(Boeing)*

General Delos C. Emmons and Major Vincent Meloy, completed an equally successful flight to Rio de Janeiro, Brazil. Aside from confirming the 2nd Bomb Group's navigational skill and airmanship, these trips reminded any would-be aggressor that the Air Corps now had the ability to fly bombers over long distances.

The wonderful achievement, for which the 2nd Bomb Group was awarded the MacKay Trophy in 1939, also demonstrated to the Army Air Corps' detractors that it now had the ability, if needed, to carry out strategic bombing. However, there were still officers in the War Department in Washington DC who continued to ignore the words of GHQ Air Force Commander General Frank Andrews who, in June 1937, had urged the War Department that all future bombers should be four-engined.

Bomber setback

In May 1938 the War Department announced that for the fiscal years 1939 and 1940 experimentation and development would be confined to aircraft designed for the close-in support of ground troops. The War Department also stated that aircraft production would be restricted to medium and light bombers, pursuit and other light aircraft. To their chagrin, the American advocates of air power

knew that while the Chiefs of Staffs continued to procure twin-engined bombers the Air Corps would, in any future war, be confined solely to a supporting role for the troops on the ground.

There were those, such as General Andrews, who, like General William 'Billy' Mitchell before him, remained convinced that wars could be won by strategic air power alone. In spite of the fact that Mitchell had been court martialled for his beliefs, they were convinced of the need for a genuine strategic bomber which could destroy America's enemies before they reached the battlefields, though it has to be said that they had not yet seen what the German Blitzkrieg could achieve.

Multi-engine engine development was stifled therefore and aircraft manufacturers were left to fight for scraps. An order for just ten B-17Bs was received by Boeing in August 1937, and only 29 more had been ordered by June 1938. To make matters worse for the company, the US Government haggled over the cost of each B-17. Boeing was a small, independent company, with barely 600 employees, and tooling up in anticipation of large production orders had left them with no cash reserves.

In May 1938 the AAC carried out a 'navigational exercise' designed to reinforce the

The first B-17B was 38-211 MD105. It was assigned to the Air Corps' Materiel Division at Wright Field on 2 August 1939 – hence the 'MD' unit designation. Apart from detail differences, the 'B' model's major innovation was the adoption of more powerful Wright Cyclone engines. *(Boeing)*

need for a strategic bombing capability, at the same time demonstrating to the US Navy that their fleets no longer retained complete mastery of the sea. Three Y1B-17s of the 49th Bomb Squadron were tasked with intercepting the Italian liner *Rex* some 725 miles off the coast of America. Olds and his navigator, Lieutenant (later General) Curtis E. LeMay, completed the task and dropped a message onto the deck of the liner. This brilliant feat of navigation proved that an invasion force at sea could be intercepted before it could harm coastal defences. The Admirals in the US Navy, so shocked that their traditional sphere of influence was under threat, lobbied the War Department. General Malin Craig, AAC Chief of Staff, was prompted to issue a policy declaration limiting the AAC's area of operation to not more than 100 miles from the US coastline!

Supercharged performance

In the summer of 1938 a 2nd Bomb Group Y1B-17 (36-157) flown by Lt William Bentley, emerged intact after being thrown onto its back in a violent thunderstorm during a flight to Langley Field. This rendered unnecessary a static test aircraft (37-369, the fourteenth Y1B-17, ordered on 12 May 1937) which was to be used in a controlled experiment to discover the maximum amount of stress the aircraft could

take before it disintegrated. Instead, 37-369 (now redesignated Y1B-17A/Boeing Model 299F and fitted with supercharged R-1820-51 (G5) engines) was used to test the Moss/General Electric turbo-superchargers needed for high altitude flight.

Supercharging offered tremendous performance advantages. The power produced by a piston engine is directly related to the amount of air passing through it in a given time, so the greater the mass of air that can be 'rammed' into the cylinders, the greater the high-altitude performance will be. Dr Sanford Moss of General Electric had developed a supercharger or 'blower', a pump which compressed and delivered higher pressure air to the engine. it was first used in flight tests in 1920. There were two types in use, the mechanically-driven supercharger in which the pump was driven by the engine crankshaft, and the turbo-supercharger, driven by engine exhaust gases.

The Y1B-17A first flew on 29 April 1938 with the supercharger turbines mounted atop the R-1820-51 engine nacelles. This had been requested by the Air Corps, which specified that the exhaust be at the top of the nacelle to simplify the installation and reduce ducting problems. However, this location, and the turbulent exhaust, disturbed the airflow over the wing and the experiment failed. With the turbos mounted under the nacelles, the Y1B-17A flew successfully on 20 November. It was to go on to set a load-carrying record by hauling 11,000 lb over a distance of 620 miles at an average speed of 238 mph.

On 31 January 1939 37-369 was delivered to the Materiel Division at Wright Field for experimental testing. With the turbo-supercharger engaged, each R-1820-51 Cyclone was capable of producing 800 hp at 25,000 feet, to give a top speed of 295 mph. The R-1820-39, which powered the original Y1B-17, generated just 775 hp at 14,000 feet giving a top speed of only 256 mph. Turbos became standard on the B-17B and all future B-17 models.

President's plea

In January 1939 Franklin D. Roosevelt asked Congress to strengthen America's air power, declaring that it was 'utterly inadequate'. The AAC pressed for more four-engined bombers, but by June 1939 there were barely thirteen B-17s operational and few more in sight. On 3 August 1937 Boeing had received a production order for only ten B-17Bs (Model 199E, later Model 299M). By 30 June 1938 orders placed had totalled just thirty-nine. Financial considerations played their part, and the US Government tried to limit expenditure by paying less for the actual articles.

Boeing had spent $100,000 on the supercharger development and wanted payment. Although the Air Corps had previously agreed to pay $205,000 per aircraft it now offered only $198,000. Boeing stuck to their guns, not least because they would actually make a loss on the aircraft, and eventually, a compromise was worked out where the Army would pay $202,500 per aircraft.

The B-17 was beset with technical problems, not least the superchargers. These were tending to fail at an alarming rate, were very sensitive to heat and cold, and would crack if not operated correctly. On occasion they burst into flame and threatened to set the wing on fire if not extinguished quickly. B-17 development lagged as a result of these problems and 38-211, the first B-17B/Model 299M (changed from 299E) did not fly until 27 June 1939.

The B-17B was powered by improved Wright R-1820-51 Cyclones that developed 1,200 hp for take off and which gave 900 hp up to 25,000 ft. The kinked forward fuselage and small rotating turret on the nose were deleted to produce a new, more streamlined nose, which was 7 inches shorter than that of the Y1B-17. The Plexiglas had a flat bomb-aiming panel with a simple socket for a .30-calibre Browning machine-gun. The crew compliment was increased to ten, with the navigator becoming a separate crewmember moving from behind the pilots to a more practical position in the new nose section, on the left-hand side behind the bombardier. The metal-skinned flaps were enlarged by moving the ends of the inner wing-panels outboard five main rib spaces, and by shortening the ailerons. The rudder was of increased area, and a Plexiglas dome was added to the cabin roof for the aircraft commander, who sat behind the pilot. External bomb racks could be added to carry a further 4,000 lb of bombs if needed. In-fuselage flotation bags were deleted, and provision was made to carry two auxiliary fuel tanks in the bomb bay. The brakes were changed from the pneumatic type of the Model 299 and Y1B-17, to hydraulic types. The B-17B first flew on 27 June 1939, the first (38-211) being retained at Wright Field to test new armament installations intended for the B-17C.

From 29 July 1939 to 30 March 1940, 39 B-17Bs were issued to the 2nd Bomb Group at Langley Field, Virginia, and to the 7th Bomb Group at Hamilton Field, San Francisco, California. In October 1939 the 2nd Bomb Group's original B-17s were transferred to the 19th Bomb Group at March Field, California.

Lessons from combat

America was at peace, but combat operations would soon produce new challenges. The B-17B's ability to reach uncharted altitudes posed new problems, not least among the crews who had to operate in very cold temperatures, while oil and other lubricants tended to take on the consistency of tar. In 1940–41, many B-17Bs were revamped and fitted with new devices such as flush-type waist windows for .50-calibre guns.

In 1940 the 2nd and 7th Bomb Groups, equipped with B-17B high altitude bombers, practised precision bombing using the top-

B-17D at Seattle on 6 December 1941. Even though the Fortress had not been particularly effective in British hands, The 'D' incorporated many design changes as a result of the experience gained by the Royal Air Force's No. 90 Squadron over Europe. Note the ventral 'bathtub' gunner's position. *(Boeing)*

secret gyro-stabilized Norden bombsight, originally developed by Carl L. Norden and Capt Frederick I. Entwistle. Experienced bombardiers placed their practice bombs within yards of the target from as high as 20,000 ft; a feat which led to claims that bombs could be place in a pickle barrel from such heights.

The B-17C (Model 299H), which flew for the first time on 21 July 1940, was a more combat-worthy model following recommendations made by Britain and France as a result of their experience with bombers in air combat. Armour plate (in the tail behind the waist positions) and self-sealing fuel tanks were fitted, and all machine-guns, except the nose-gun, were standardized at .50-calibre. The two limited-vision gun cupolas on the sides of the fuselage were replaced with streamlined, Plexiglas, teardrop-shaped flush windows, and the guns moved inside onto swivel posts.

Combat experience in Europe had also revealed a need for all-round defence, and this could only be addressed by installing power-operated gun turrets with belt-fed guns – aboard the B-l7C all six guns were pannier-fed. However, the top gun blister was replaced only by a removable sliding Plexiglas hatch, while a large 'bath-tub', in which the gunner had to kneel to fire his gun, replaced the under gunner's blister. Both only provided for rearward defence.

Good performance

Nevertheless, the B-17C was considered well armed, and it possessed an impressive top speed of 325 mph at around 29,000 ft and could cruise at 230 mph at 30,000 ft. The bomb load remained the same (4,996 lb, as on the B-17B). Other changes included adding boost and transfer pumps to allow each fuel tank to feed a separate engine, and the oxygen system was changed to a manifold type. Dual brakes replaced single units on each main wheel. Thirty-eight B-17Cs were ordered on 10 August 1939, the first being retained by Boeing for test purposes.

In 1939 the overall US requirement was for 461 new B-17, B-24, B-25 and B-26 bombers. Only the B-17 had flown in prototype form and Britain acquired twenty of these new aircraft (Model 299U) as the Fortress I. By September 1941 the RAF had decided that the Fortress was unsuitable for further operations with Bomber Command. Although it was an extremely well built aircraft, operational experience revealed that its defensive firepower of five 0.3-in and

A Fortress I (40-2064) which became AN528 *B-Baker* in No. 90 Squadron RAF pictured during an early test flight in Washington State. *B-Baker* joined No. 90 Squadron on 4 June 1941 but it caught fire running up its engines at dispersal at Polebrook on 3 July and was burnt out. *(Boeing)*

0.5-in machine-guns was totally inadequate for operations over heavily defended targets in Europe. The RAF operated the Fortress I at altitudes approaching 30,000 ft. The top secret Norden bombsight had not been sold to the British, and was replaced by the Sperry sight. As a result, in 26 out of 51 sorties no bombs were dropped: bomb aimers did not feel able to hit the target accurately.

Meanwhile, in America, the 1940 requirement for B-17, B-24, B-25 and B-26 bombers had risen to 3,214, but orders were placed piecemeal – by the summer of 1941 only 700 bombers of all types were in service, and companies like Boeing were struggling to stay afloat. In April 1940 the Army had finally exercised its option for 42 more B-l7Cs, but only

after Boeing had been forced to lay off part of its workforce. On 18 May Boeing received a contract for 240 DB-7 attack bombers for France, to be built under licence from Douglas. France capitulated in June 1940 before the contract was completed and the aircraft subsequently were taken over by Great Britain.

On 12 July 1940 the War Department advised Boeing that orders for 512 more B-17s in two lots (277 and 235) would soon follow. The B-l7Cs required so many modifications as a result of the RAF experience that they were redesignated B-l7Ds on 6 September 1941. Outwardly, the 'D' differed from the 'C' by the addition of engine cowl flaps and twin-gun installations in the belly and upper positions. Internally, more armour plate was added, a

Fortress I AN529 (B-17C 40-2065) *C-Charlie* at Blackpool, in April 1941, shortly after arrival from the USA. It was one of the first B-17Cs taken on charge by No. 90 Squadron RAF in 1941. AN529 was lost in the Libyan Desert on 8 November 1941, during a daylight raid on Benghazi. *(via Mike Bailey)*

new self-sealing fuel tank system installed and changes were also made to the bomb release, oxygen supply and to the aircraft's electrical systems. In May 1941 21 B-17Ds were flown to Hawaii and the remainder were sent to the 7th and 19th Bomb Groups.

Army Air Force

On 21 July 1941 the Army Air Forces (AAF) were created, under the command of Major-General Henry H. 'Hap' Arnold. In the same month, President Roosevelt asked the Secretaries of War and of the Navy to produce estimates for bringing their forces to an effective war footing. Arnold used the opportunity to gain permission for the AAC's Air War Plans Division to prepare their own report, forcing the War Plans Division to concentrate solely on the needs of its land forces. Arnold's staff officers at AWPD, headed by Col Harold L. George, formulated a policy (AWPD/1) of a relentless air offensive against Nazi Germany and a strategic defence in the Pacific. If Japan entered the war, it too would be subjected to aerial bombardment, but only after Germany had been beaten.

AWPD/1 calculated that six months of strategic bombing using daylight precision attacks on Nazi industry, together with the neutralization of the *Luftwaffe*, submarine and naval facilities, might render a land campaign against Germany unnecessary. They would need to destroy airframe assembly plants and associated metal production, and some more

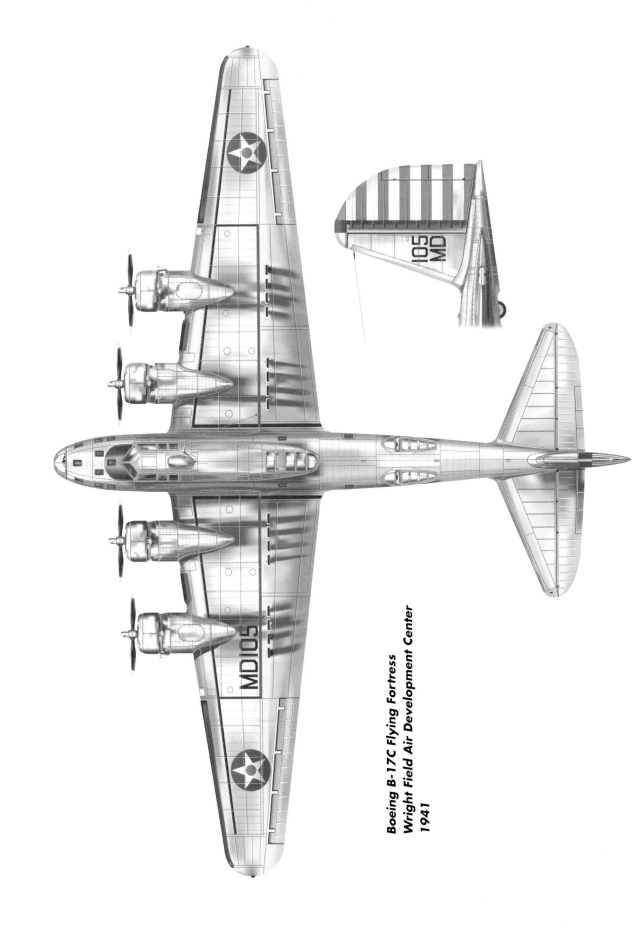

Boeing B-17C Flying Fortress
Wright Field Air Development Center
1941

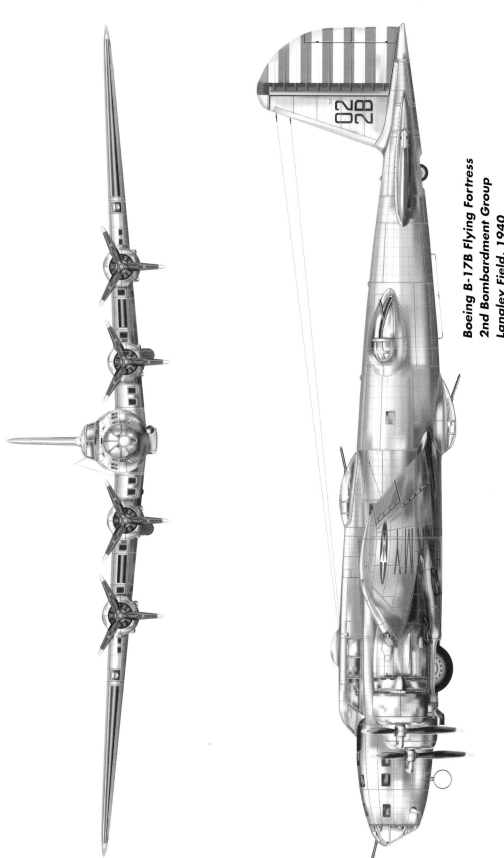

**Boeing B-17B Flying Fortress
2nd Bombardment Group
Langley Field, 1940**

The early B-17 proved to be quite inadequate for combat against modern fighter defences. The next model, the B-17E, would be radically redesigned to make the big bomber more combat ready. This B-17E, 41-2600 *Esmeralda*, is being used for training at Ephrata, Washington, in the summer of 1942. In the background are B-17E 41-2557 and a B-17C. *(Boeing)*

specialized targets as well. Fifty electrical generating or switching stations, 47 key points in the German transport network, and all of the 27 petroleum plants in Germany – 154 targets in total – was considered achievable if 90 per cent of all bombs dropped exploded within 1,250 feet of the Mean Point of Impact (MPI).

Plans for war

To achieve its aims, the AAF would need good weather on a regular basis in European skies, and a massive force of continental and intercontinental bombers. These would have to number somewhere in the order of 1,060 medium bombers, 1,700 B-17 and B-24 heavy bombers (in twenty groups), 2,040 of the planned B-29 and B-32 very heavy bombers and 3,412 fighters. And all of these would need to be deployed against Nazi Germany from bases in Great Britain and Egypt. Additionally, another 3,740 intercontinental bombers – the Consolidated B-36 was then being proposed – would, it was thought, strike from bases in the

US. However, American industry did not yet have the capacity to build bombers in such numbers. Some types, like the B-32 never entered full-scale production, and the sheer enormity of the B-36 project prevented large-scale production until after the war.

Sixteen pursuit groups were thought necessary to protect the bomber bases, but because the planners believed that the bombers could fight their way to a target and back again without prohibitive losses, no provision was made for long-range escort fighters to accompany them. Unfortunately for the crews of the B-17 Flying Fortress, and to a lesser extent other types, it was their lot to try to make the theory of daylight precision bombing work, without adequate fighter cover or the support of the larger and longer ranging very heavy bomber types. While the press, newsreels, and even Hollywood movies championed Boeing's gleaming new Flying Fortresses in glowing terms, reality would be quite different.

2. Operational History 1: Britain, the Pacific and the Med

The first B-17s to fight were the twenty Fortress Is (B-17Cs) serial numbered AN518 to AN537, of No. 90 Squadron RAF, which began bombing operations from England in the spring of 1941. Most of the aircrew personnel assigned to the squadron were veterans of RAF night bombing or low-level daylight operations. Because the Fortress I was expected to fly at heights well in excess of 30,000 ft, an altitude far beyond operational RAF bombers (except for the pressurized experimental Wellington Mk V/VI), all of the young airmen had to first undergo a rigorous decompression test at Farnborough. This consisted of 'climbing' at 3,000 ft/min to 35,000 ft, and remaining there for 5 hours.

The B-17C had an impressive top speed of 325 mph at around 29,000 ft and could cruise at 230 mph at 30,000 ft. However, the only armour plate was fitted in the tail behind the waist positions, and there were no power operated gun turrets. Even so, the B-17C was considered well armed, with one or two .30-in machine-guns in the nose and six .50-in pannier-fed guns – two in the waist, two in a ventral cupola and two in the dorsal position.

Range was poor however, and only American bombs, up to 1,000 lb, could be carried. The AAC had not been willing to export the top secret Norden precision bombsight, and it was replaced on British machines by the Sperry sight which was only calibrated for automatic operation to 25,000 ft. At higher altitudes RAF bomb aimers had to 'guesstimate' by feeding in pre-set calculations supplied by Sperry. Despite these limitations the RAF need for operational bomber aircraft was so great that once the aircraft had been fitted out and the crews trained, they were used by No. 2 Group for bombing missions.

High altitude problems

The Army Air Corps provided five experienced airmen to help the British, while other American advisors included an expert on the Sperry 01 bombsight, a number of Boeing representatives and an expert on Wright Cyclones and superchargers. There were many problems with the oxygen and intercom systems, which needed rectifying before the Fortress I could be operated at altitude, as Flying Officer Antony J. Barwood, a young medical officer, recalls.

'We started with the American Oxygen system, A8 individually controlled regulators and BLB re-breather bag masks with hand-held carbon granule microphones. The regulators seized up, the masks froze and the microphones became progressively more useless above 15,000 ft as they depended on air density to excite the carbon granules within the diaphragm of the microphone. We then changed to British Mk VIII oxygen regulators and Type "E" masks with an incorporated electromagnetic microphone, which also required amplified changes in the aircraft. The masks still froze and were modified with an additional valve. I covered the diaphragm inside the microphone with a French letter to prevent it from freezing.'

'An oxygen economiser, which had been invented by Professor Brian Matthews – later Sir Brian – was introduced. It stored the oxygen flowing through the regulator while the user was not breathing in, which is only about one-third of the breathing cycle. The original economisers were hand-made by "metal bashers" at the Royal Aircraft Establishment and at the Physiology Laboratory. They effectively reduced the weight of oxygen cylinders which the aircraft had to carry by 50 per cent, and produced a more effective oxygen system. The final change was to a Mk 10 regulator controlled centrally by the captain delivering oxygen to each crew position.'

Fighting the cold

The bitter cold at altitude was made far worse by the aircraft having to fly with all four of the rear fuselage blisters off to enable the gunners to fire. That high air blast affected all the rear crew, making effective flying clothing of vital importance. Barwood again:

'At first we used electrically heated one-piece suits made by Seibe-Gorman, with electrically heated gloves and boots. The suits restricted movement, which was so essential for the gunners. They were bulky and not very reliable. In August the "Taylor" suit became available, again one-piece but much more easily donned, with an electrically heated lining, glove lining and socks. These were used with fleecy-lined flying boots and soft leather gauntlets. The suit also provided built-in flotation. It was reliable and much easier to move in.'

There were technical problems to contend with, too, as one of the gunners, Sgt Tom Imrie DFM, recalls:

'We had constant engine oil problems caused by the pressure differences. The oxygen system and the intercom were bad. Armament was prehistoric, with free-mounted .50s in the waist and one .30 in the nose. Ammunition was contained in heavy 50 lb containers and it was a hell of a struggle trying to lift them onto the mountings at 30,000 feet plus. The guns jumped around all over the place, and hosepiped on the free mountings. Often they didn't fire. They iced up at altitude and we had to wash them in petrol. The windscreens iced up too and

eventually had to be double-glazed. It was nerve wracking flying in broad daylight. On one test flight, over Cornwall, on 4 June 1941, we even got the Fortress up to 41,000 feet. We could see the earth's curvature and the sky had turned a dark purple colour instead of blue.'

During training, one of the RAF Forts hit a layer of cumulonimbus at about 30,000 feet, fell 12,000 feet and broke up. There was only one survivor. A second was lost when it caught fire on the ground during an engine test. The first operation went ahead on 8 July 1941 when three Fortress Is took off from Polebrook and attacked Wilhelmshaven. Each carried four 1,000-lb ground burst bombs, since armour–piercing weapons were not yet available. Results were unobserved.

By August apprehension was growing about the B-17C's ability to remain immune from attack at high altitude, but operations continued. In September 1941 four Fortresses tried to bomb the pocket battleship *Admiral von Scheer* in Oslo fiord. Two Fortresses were lost, one aborted, and a fourth just managed to make it home after being shot up by fighters. Only four more individual sorties were flown after the Norway debacle and to all intents and purposes, No. 90 Squadron's brief association with the Fortress was over. In 51 sorties by individual aircraft on 26 raids, 26 were abortive and no bombs were dropped.

On 26 October four Fortress Is were flown to the Middle East for daylight raids on Benghazi and Italian shipping. Two of the Fortresses were lost and the two surviving machines were flown to India in February 1942. They were handed over to the USAAF in December that year for training, ending British operations with early B-17s.

B-17s in the Pacific

At the time of Pearl Harbor, Sunday 7 December 1941, when Japanese carrier-borne aircraft made a pre-emptive strike on the US Fleet in the Hawaiian Islands, a few B-17Ds were based in Hawaii and more were en route. B-17Ds of the 5th and 11th Bomb Groups were among the many aircraft lined up in neat rows that were destroyed on the ground. Twelve unarmed B-17Ds of the 7th Bomb Group, and

B-17Es 41-2459 and 41-2461 caused immense interest among the British crews of No. 90 Squadron when they passed through the desert airstrip at Shallufa in December 1941, en route to join the 19th Bomb Group in the Pacific. The British, who had been flying Fortress Is (B-17Cs) were particularly interested in the new tail design and rear armament. The 19th fought a gallant, but losing battle in the Philippines after the Japanese invasion and were forced to retreat to Australia on Christmas Eve 1941, before returning, briefly, to Java late in the month. Within two days of their arrival on 16 January 1942, 41-2459 and 41-2461 were despatched on a mission with three LB-30 Liberators. Lt. J. L 'Duke' Du Frane's crew in 41-2459 force-landed at Kandari, Borneo and were strafed by Zeros. The crew escaped and were later evacuated to Java. The aircraft was later blown up during the US retreat. 41-2461, piloted by Major C. R. Nacrasson was hit but got away safely. *(Antony Barwood)*

A Japanese-captured B-17E in flight, photographed from a B-17D that had also been captured. Altogether, the Japanese obtained three Fortresses, two B-17Ds and one early B-17E, which were flown to Japan and put on public display with other captured US aircraft. These captured Fortresses were carefully evaluated and were used to develop fighter tactics against them. *(USAF)*

four B-17Cs and two B-17Es of the 88th Reconnaissance Squadron, all of which were en route to Mindanao in the Philippines, flew in from Hamilton Field, arriving over Hawaii during the Japanese attack. Pilots landed wherever they could. One pilot put down on a golf course, while some landed on the small fighter strip at Bellows Field.

US bomber strength in the Pacific

At the time of the Japanese attacks, America had some thirteen groups equipped with the B-17, but most were well below the group strength of thirty-two aircraft. Some 150 B-17s, of all models and including twelve YB-17s, were scattered throughout the Pacific seaboard, Alaska and Newfoundland. Twenty-nine B-17Es of the 7th Bomb Group, which left Salt Lake City, Utah, on 5 December for the

Philippines, were urgently diverted to Muroc to help defend California from possible enemy attack. Nineteen B-17Bs were despatched to Spokane, Washington to join the five B-17Cs of the 19th Bomb Group, while just two B-17Bs were stationed in Alaska. Six B-17Bs of the 41st Reconnaissance Squadron were stationed in Newfoundland. During the first week of December, eight B-17Bs and nineteen B-18s were delivered to the 6th Bomb Group, which had arrived at Rio Hato, Panama, on 9 December, to defend the Panama Canal.

In US service the B-17 Flying Fortress would serve in three main theatres of war – in the Southwest Pacific, the South Pacific, and the ETO (European Theatre of Operations). In the Southwest Pacific B-17s served in the 19th and 43rd Bomb Groups, V Bomber Command, 5th Air Force, and in the 11th and 5th Bomb Groups

B-17E 41-2435 of the 40th Bomb Squadron, 19th Bomb Group over the Owen Stanley mountain range in 1942. This aircraft was shot down off Buna on 12 August 1942. The 19th Bomb Group left Australia for the US on 1 November 1942. Note the impractical remote-controlled Bendix ventral turret, which was operated by a gunner lying prone to its rear and sighting through a system using mirrors contained in a small transparent blister. Two low slit windows on both sides of the fuselage were located for the gunner to identify his target. *(Ken Fields)*

of the 13th Air Force. The 5th and 13th Air Forces used Fortresses only until mid-1943 because B-17s were more urgently needed in the ETO. In any case, the B-24 Liberator was more suited to the long distances involved in the Pacific theatres of war.

Combat against Japan

The 19th Bomb Group returned to the USA late in 1942 and became a replacement training unit before being re-activated on 1 April 1944 as a B-29 Superfortress group. The 43rd Bomb Group converted to Liberators between May and September 1943, and the 11th Bomb Group transferred to the 7th AF (Air Force) in March 1943 to retrain as a B-24 unit. The 5th Bomb Group converted to B-24s in December 1943. Though the B-17 served for only a relatively short time in the vast Pacific, often in small numbers, it nevertheless played an important part in trying to stem the tide of early Japanese dominance before the decisive naval battles at the Coral Sea and at Midway.

The outbreak of war caught the Far East Air Force units mostly on the ground, just as they had been in Hawaii. At the time of the attack there were about 35 B-17C and Ds of the 19th Bomb Group at Clark Field, Luzon. Most escaped by getting airborne, but when they landed back at Clark and Iba to refuel another attack destroyed eighteen B-17s on the ground.

Only one B-17 at Clark Field, and sixteen B-17Cs of the 14th Bomb Squadron at Del Monte, a small satellite field on Mindanao, some 600 miles to the south of Clark, escaped. On 9 December the 19th Bomb Group flew limited reconnaissance patrols in a vain search for the Japanese invasion fleet.

The crew of *Li'l Nell* of the 5th Bomb Group at Henderson Field, Guadalcanal, 15 August 1943. Back row, L–R: S/Sgt Hildebrand, engineer; Capt A. D. Lucas, pilot; Lt W.L Chestnut, co-pilot; Lt Alfred B. Cohen, navigator; Lt W.A. Hodges, bombardier. Front row: Cpl Waselowski asst engineer; Cpl Granowski, asst radio man; Sgt Hamaker, tail gunner; Sgt Fredricks, radio operator (KIA Sept 1943). The censor has obliterated the fuselage aerial. *(Alfred B. Cohen)*

On 10 December the 19th Bomb Group carried out the first American bombing raid of the war when they made an abortive attack on a Japanese convoy landing troops and equipment at Vigan and at Aparri in northern Luzon. Two days later some of the of the B-17s were flown 1,500 miles south to Batchelor Field, Darwin, in northern Australia, from where small and sporadic attacks were made on Japanese shipping at Davao Bay, Mindanao, and at Lingayen Gulf, Luzon.

All AAF units on Luzon, as well as ground forces, began leaving for the Bataan peninsula. The 19th Bomb Group air echelon was abandoned and the ground personnel were sent to bolster the infantry. Late in December 1941 the air echelon moved to Australia to transport medical and other supplies to the Philippines and evacuate personnel from the area. The 19th

Bomb Group moved to Java at the end of 1941 and, flying B-17 and LB-30/B-24 Liberator aircraft, earned a Distinguished Unit Citation or DUC for attacking enemy aircraft, ground installations, warships, and transports during the Japanese drive through the Philippines and Netherlands East Indies early in 1942. The 19th Bomb Group returned to Australia from Java early in March 1942, and later that month the group evacuated General Douglas MacArthur, his family, and key members of his staff from the Philippines to Australia.

SW Pacific battles

By mid-April 1942, the Japanese were well on the way to total domination in the New Guinea-New Britain-Solomon Islands area of the South Pacific. The turning point however, came when the Japanese invasion fleet heading for Port Moresby was turned back in the Battle of the Coral Sea on 7–9 May; the first battle in history in which the two naval forces did not exchange fire but which was decided by the two air fleets.

The 19th Bomb Group took part in several actions, and later hit Japanese transportation, communications, and ground forces during the enemy's invasion of New Guinea. From 7 to 12 August 1942, the 19th Bomb Group bombed aerodromes, ground installations, and shipping near Rabaul, New Britain, being awarded another DUC for these missions. The group returned to the US late in 1942.

On 3 June Japanese forces attacked Midway Island. A handful of B-17Es of the 11th Bomb Group made several reconnaissance flights from Midway, and once combat began made a number of bombing runs against the Japanese invasion fleet. Between 3 and 5 June the B-17Es flew sixteen attacks, a total of 55 sorties, for the loss of two aircraft. Although B-17 crews claimed several strikes on 4 and 5 June, none of the enemy ships were in fact hit. Nevertheless, the battle ended in victory for the US, in spite of heavy losses. Four of Japan's most powerful carriers had been sunk, taking the cream of the Imperial Navy's aircrews with them.

In August 1942 the 43rd Bomb Group at Port Moresby joined the 5th Air Force. The 43rd Bomb Group had been activated on 15 January 1941 and had moved to the southwest Pacific

General Douglas MacArthur's B-17E-BO (XC-108) 41-2593 *Bataan*, one of four B-17s specially converted into transports under the C-108 designation in 1943. Although an 'E', *Bataan* had a B-17F Plexiglas nose with a single .50 calibre machine-gun with chromed barrel (the only armament carried), and a navigator's astrodome added to the aircraft. A five-man crew and up to eleven passengers could be carried on the XC-108. *(Boeing)*

via Cape Town, South Africa, during February and March 1942. Between October 1942 and February 1943, several day and night attacks were made on Rabaul on New Britain, the main Japanese base in the Pacific. Some crews flew as low as 250 feet to hit their targets, and in a raid on shipping in Rabaul harbour on 5 January, Brig-Gen Kenneth Walker, commanding general, 5th Bomber Command, was killed aboard one of the two B-17s shot down.

Anti-shipping strike

The 43rd Bomb Group experimented with 'skip'-bombing – releasing bombs at low level and bouncing them off the surface of the sea into the side of target vessels – and used this method for shipping strikes, including several decisive actions during the Battle of the Bismarck Sea, 2–4 March 1943. The 43rd Bomb Group received a DUC for participation in this battle, which annihilated Japanese forces attempting to reinforce New Guinea. Other operations during this period included support for ground forces on New Guinea; attacks on airfields and installations in New Guinea, the Bismarck Archipelago, Celebes, Halmahera, Yap, Palau, and the southern Philippines; and long-range raids against oil refineries on Ceram and Borneo.

In May the 43rd Bomb Group began converting to the B-24, and by the end of September had fully converted to the Liberator. Meanwhile, late in July 1942 B-17s of the 11th

Bomb Group arrived at New Caledonia for operations in the 13th Air Force against Guadalcanal. The 'Canal' was a hilly, tropical, jungle-covered island in the Solomon Islands group, where on 4 July the Japanese had started building an airfield from which to threaten the Allied lifeline to Australia. Between July and November 1942, the 11th Bomb Group attacked enemy airfields, supply dumps, shipping and troop movements, receiving a DUC for its actions. From late 1942 to late March 1943 the 11th Bomb Group carried out attacks on enemy airfields, and shipping in the Solomons before returning to Hawaii for reassignment to the 7th AF and re-equipment with B-24 Liberators.

The 5th Bomb Group, meanwhile, had left Hawaii in November 1942 and joined the 13th Air Force, participating in the Allied drive from the Solomons to the Philippines. Up until August 1943 they flew long patrol and photographic reconnaissance missions over the Solomons and the Coral Sea, attacking Japanese shipping off Guadalcanal as well as raiding airfields in the northern Solomons. Many enemy bases, including Bougainville, New Britain, and New Ireland, were bombed. The Group converted to B-24 Liberators in December 1943.

B-17 Flying Fortresses were badly needed in Europe, where they equipped many groups in the 8th AF in England. However, they also played a major part in operations in the

B-17E 41-2400 was built by Boeing and delivered to Douglas as the model for B-17F production. It was converted into a transport after many months of combat in the South Pacific. *Oklahoma Gal* had an extremely impressive combat record: she flew 203 bombing missions, sinking eight Japanese ships, and her gunners were credited with shooting down six enemy aircraft. *(Douglas)*

Mediterranean. Air forces in that theatre underwent many changes of designation. The 12th Air Force, supporting Operation *Torch*, the invasion of North Africa, was actually created by transferring fourteen fighter, bomber and troop carrier units from the 8th Air Force to the new air force in September and October 1942.

African invasion

Operation *Torch* went ahead on 8 November 1942, with landings at Casablanca, Oran and Algiers. Two B-17F groups, the 97th and 301st Bomb Groups, which had previously flown a handful of missions from eastern England, made up the 5th Bomb Wing, 12th Air Force. On 16 November 1942 the 97th Bomb Group, which had flown the first 8th Air Force B-17 mission from England on 17 August 1942, flew the first mission by the 12th, when six of its B-17s raided Sidi Ahmed airfield at Bizerte,

Tunisia. On 22 November the B-17s were moved back from Maison Blanche near Algiers, to Tafaraoui, Algeria, because of Axis bombing raids on their airfield. On 28 November, 35 B-17s of the 301st and 97th Bomb Groups raided Bizerte airfield and dock area, losing two B-17s.

From December 1942 to the end of January 1943, bombing raids were made almost daily on key Axis targets in Tunisia and Tripolitania. In February 1943, when the B-17s were assigned to the Northwest African Strategic Air Force (NASAF), the 5th Bomb Wing was joined by the 99th Bomb Group. In March 1943 the 2nd Bomb Group arrived. In July 1943 the B-17s took part in Operation *Husky*, the invasion of Sicily. The 99th Bomb Group received a DUC for its performance on 5 July, when it helped neutralize fighter opposition prior to the invasion by penetrating enemy air defences to bomb Gerbani airfield.

General Henry H. Arnold's plan to create a Strategic Air Force in the Mediterranean was accepted: he proposed splitting the 12th in half, leaving the other half a tactical organization. On 1 November 1943, the 15th AF was officially activated in Tunis under the command of Major-General Jimmy Doolittle. Four of the groups were equipped with B-17s. The first 15th AF bombing raid of the war was on 1 November when the 5th Bomb Wing attacked the La Spezia naval base and the Vezzano railway bridge nearby. The next day, 74 B-17s hit Wiener-Neustadt, in Austria, losing six Fortresses. Further raids were made on targets in northern Italy, Greece, Bulgaria and France, though bad weather restricted operations during November and December.

Italian bases

In December 1943 and January 1944, the four B-17 groups moved to bases in the Foggia area of southern Italy. Early in 1944 the 463rd Bomb Group and the 483rd Bomb Group brought the number of B-17 groups in the 5th Wing to six.

New Year 1944 saw the 15th AF participate in combined bomber offensive missions in support of the *Pointblank* offensive against German aircraft plants, ball-bearing factories, oil refineries, and other targets in Germany. The 15th AF took part in Operation *Argument*, a concerted series of co-ordinated daylight precision raids on enemy aircraft and airframe assembly plants.

The first of the raids, which would become known as 'Big Week', began on 20 February, when 126 B-17s of the 15th AF had to abort their mission to the Bf 109 assembly plant at Regensburg because of severe icing conditions over the Alps. On 21 February the 15th was grounded by bad weather in the Foggia area. On 22 February 65 B-17s of the 15th AF were despatched to Regensburg. Meanwhile, one group of 21 B-17s dropped 42 tons of bombs on the marshalling yards at Peterhausen, and a further 28 bombed an airfield at Zagreb, Yugoslavia. Five B-17s were shot down.

On 24 February B-17s led by the 97th Bomb Group raided the Steyr-Daimler-Puch aircraft plant. Despite a heavy escort of over 140 P-38s and P-47s, enemy fighters shot down 14 B-17s of the 2nd Bomb Group bringing up the rear.

B-17G-50-DL 44-6397 of the 416th Bomb Squadron, 99th Bomb Group, drops its bombs on the Szob railway bridge, north of Budapest, on 30 March 1944. Initially, all the 463rd Bomb Group's B-17Gs which crews had flown into theatre were given to the 99th Bomb Group, in return for their older and many war-weary B-17Fs. After crews had flown in combat with other groups, the 463rd Bomb Group flew its first mission on 30 March 1944. 44-6397 was lost on 23 March 1945, on a mission to Ruhrland. *(USAF)*

Cloaked in secrecy, 130 B-17s of the 2nd, 97th, 99th and 483rd Bomb Groups flew the first 15th Air Force 'shuttle' mission on 2 June 1944. Led by Lt-Gen Ira C. Eaker, the mission, codenamed *Frantic Joe*, was in support of Russian operations in the Balkans. Seen here on 11 June, B-17Gs of the 49th Bomb Squadron, 2nd Bomb Group, are heading from Russia to bomb Foscani airfield in Romania, during the return trip to Italy. *(USAF)*

Total B-17 losses this day were 16. A smaller force of 27 B-17s became separated and bombed the oil refinery and torpedo works at Fiume for the loss of only one B-17.

Strategic campaign

On 25 February the USSTAF despatched 1,154 bombers and 1,000 fighters to bomb aircraft plants, ball bearing works and components factories. Very considerable damage was caused to the Bf 109 plants at Regensburg-Prufening. Forty-six B-17s were despatched from Italy, led by ten Fortresses of the 2nd Bomb Group. Ten of the 31 B-17s of the 301st Bomb Group had to return early, leaving only 36 unescorted B-17s to continue to the target. Attacks by the *Luftwaffe* were heavy, and eleven B-17s of the 301st Bomb Group were shot down. The 2nd Bomb Group lost three.

Bad weather then grounded missions until 2 March, when close to 300 B-17s and B-24s, escorted by more than 150 fighters, made bombing strikes on enemy troop concentrations in support of the US 5th Army at Anzio. Bad weather interfered with successive missions. Then, on 7 March, about 300 B-17s and B-24s bombed Toulon, and marshalling yards and airfields in Italy, and on the 15th more than 300 B-17s and B-24s bombed Monte Cassino on the Gustav Line. The monastery atop the mountain was reduced to rubble, but the ground troops remained bogged down. On 18 March, 592 bombers and fighters – the largest 15th AF bomber formation yet despatched – bombed *Luftwaffe* airfields in Italy, with excellent results. On 30 March, 114 B-17s bombed the Sofia marshalling yards.

Communications attacks

On 2 April 1944, the 15th AF carried out the first of 26 raids on rail and transportation centres in the Balkans. On 23 April 171 B-17s attacked the Messerschmitt Bf 109 plant at Wiener-Neustadt again. Thirty-one of the 99th Bomb Group's 36 B-17s were damaged by flak and fighter attacks, and two B-17s were lost. On 24 April, five B-17s of the 301st Bomb Group, equipped with *Azon* remotely guided bombs,

B-17G-25-DL 42-38069 and other B-17Gs of the 49th Bomb Squadron, 2nd Bomb Group, are seen returning to Italy from Blechhammer, Germany on 7 July 1944. They are accompanied by a P-38 Lightning which has been hit by flak over the target and has its starboard engine feathered. 42-38069 was lost on a mission to Odertal, Germany, less than a month later, on 2 August 1944. *(USAF)*

made the first of two such attacks on the Ancona-Rimini railway line. A second raid was made on the same railway on 29 April.

In April 1944, overall command of the combined bomber offensive officially passed to General Dwight D. Eisenhower, newly appointed Supreme Allied Commander, Allied Expeditionary Forces (SHAEF). Eisenhower immediately ordered all-out attacks on German oil production centres as part of the overall plan for the invasion of Europe, scheduled for the summer of 1944. Ploesti was bombed on 15 and 24 April, and four more heavy raids followed in May.

For three days, starting on 25 May, the 15th AF bombed marshalling yards at Lyon, St Etienne, Avignon, Nimes and Marseilles. Throughout the rest of May 1944 it continued heavy raids on transportation and aircraft production targets, and bombed railway networks in southeast Europe in support of Russian military operations in Romania.

Amidst great secrecy on 2 June, the 15th AF flew its first 'shuttle' mission, codenamed *Frantic Joe*, in support of Soviet operations in the Balkans. The 130 B-17s, led by Lt-Gen Ira C. Eaker and escorted by P-38 Lightnings, bombed the marshalling yards at Debrecen, Hungary, before flying on to Russian bases at Poltava and Mirgorod. The B-17s flew a mission to Galati airfield, Romania, on 6 June, escorted all the way by P-51 Mustangs. On 11 June, the shuttle force took off from Poltava bound for Italy, bombing Foscani airfield in Romania en route.

On 23 June, the 15th began a series of raids on marshalling yards and oil targets in the Balkans, with strikes by 335 heavies on targets at Craiova, Piatra and two oil refineries at Ploesti. More than 100 aircraft were lost. On 26

A B-17G of the 347th Bomb Squadron, 99th Bomb Group, flies over a column of smoke during a heavy raid on the Ploesti oilfields on 15 July 1944. The smoke rose to a height of 23,000 ft. *(USAF)*

June, 677 B-17s and B-24s attacked targets near Vienna, and 30 aircraft, mostly B-17s, were shot down. On 7 July more than 560 B-17s and B-24s attacked two synthetic oil plants at Blechhammer, losing 18 aircraft in the process. On 9 July the Ploesti refineries were bombed for the ninth time in the first Pathfinder-led mission from Italy.

The 15th AF then returned to attacks on marshalling yards, later switching to airfields and aircraft factories in Austria and Germany. On 18 July, 167 B-17s set out for a Bf 110 and Me 410 repair and assembly installation at Memmingen airfield in Germany. Over the Adriatic, flying through thick cloud, groups became separated and 44 B-17s aborted, while 27 Fortresses attacked a secondary target. The bad weather prevented the fighter escort from rendezvousing with the remaining Fortresses. Twenty-six B-17s of the 483rd Bomb Group, led by Captain Louis I. Seith, CO of the 840th Squadron, pressed on to the primary, alone and

unescorted. About 75 *Luftwaffe* single-engined fighters ripped through the formation and shot down fourteen B-17s.

When the weather permitted, the 15th Air Force, together with the 12th Air Force, 'softened up' targets in southern France in preparation for the invasion codenamed *Anvil*. On 13 August, 136 B-17s bombed gun positions and bridges around Toulon, and on the 14th, the 15th Air Force repeated the raids with strikes by 540 B-17s and B-24s as the *Dragoon* invasion convoy headed for the French Mediterranean coast.

The heavies pounded the invasion beaches around Cannes and Toulon, while twenty-eight escorted B-17s bombed road bridges over the Rhône. A proposed strike by B-17s on coastal gun positions had to be aborted because of poor visibility in the target area. On the 16th, 108 B-17s supporting *Dragoon* attacked railway bridges near the invasion beaches.

End of the Ploesti saga

On 19 August the 15th AF bombed Ploesti for the twentieth and final time, the total including one raid by P-38s. There were no bomber losses in this last mission. Almost 13,500 tons of bombs had been dropped on the refinery complex in around two years, at a cost of 223 aircraft. However, the sacrifice ensured that production at Ploesti had been reduced to just one-fifth of its potential capacity. The Red Army finally overran Ploesti on 30 August.

Operation *Reunion* began on 31 August, when 36 B-17s, in the first of three airlifts, evacuated liberated American PoWs from Bucharest to Italy.

In early September, with the German army in headlong retreat, the 15th AF attacked the main withdrawal route across the Balkans. Attacks were made on airfields in Greece and the Aegean Islands too, when the *Luftwaffe* attempted to use 120 Ju 52s and other transports to airlift their forces. By the end of the month most of Romania and Bulgaria were under Red Army control, and by the end of the year German forces in the Balkans held only parts of Hungary and north and central Yugoslavia. Raids were mainly aimed at oil and communications targets in Austria, Czechoslovakia, Germany, Hungary and Italy,

Red-finned 15th Air Force B-17Gs of the 840th Bomb Squadron, 483rd Bomb Group, at Tri Duby (Three Oaks) airfield on 17 September 1944. Escorted by P-51B Mustangs, the Fortresses were tasked with bringing in supplies for the Slovak National Uprising. The nearest aircraft is B-17G-35-VE 42-97917, which crashed on take-off for a mission to Blechhammer on 17 October 1944. *(Zdenek Hurt)*

and other targets were bombed too. By the end of November the 15th Air Force had dropped 27,000 tons of bombs on oil targets.

These raids were costly in aircraft and crews. Apart from the threat posed by enemy fighters and Flak, the inhospitable Alps had to be crossed, and bad weather, too, could hamper or even curtail operations.

Evil weather

January 1945 marked a low point in 15th AF operations, as bad weather scrubbed most of the missions planned that month. Conditions in February were much improved, however, and the heavies began a series of twenty consecutive missions, starting on the 13th. There was no let-up during February and March, when attacks continued to be made on the enemy oil industry, but raids were mostly against German and Austrian lines of communication.

On 15 March, in the deepest penetration so far, 109 B-17s hit the synthetic oil plant at Ruhrland, just south of Berlin. The next day, more than 720 heavies pounded oil refineries again, while some B-17s and B-24s parachuted supplies to northern Italy and Yugoslavia. Bad weather prevented heavy bombing missions for two days. Then on the 19th, transportation targets in and around Vienna were hit, as more than 800 B-17s and B-24s attacked marshalling yards at several strategic points. On 22 March, 136 B-17s – part of a force of more than 680 heavies in action this day – bombed the Ruhrland oil refinery (and Lauta aluminium works to the north) and caused severe damage. Three B-17s were shot down by Me 262 jets.

On 24 March Berlin was bombed in the first 15th AF raid on the German capital, when more than 150 Fortresses set out for the Daimler-Benz tank engine plant, a 1,500-mile round trip. West of Brux the leading elements came under an

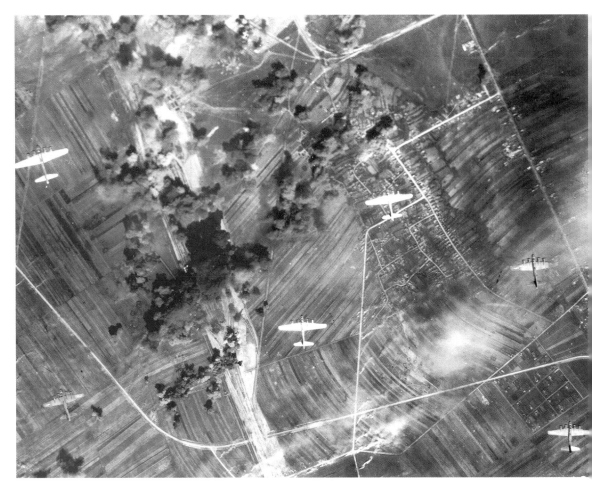

Six B-17s of the 15th Air Force bombing the railway marshalling yards at Ploesti, Romania. *(USAF)*

intense flak barrage. Four B-17s of the 463rd Bomb Group were shot down, and two more were so badly damaged that they had to return to base. Shortly after, the 463rd Bomb Group were bounced by 15 Me 262s, which shot down one B-17 and another from the 483rd Bomb Group. Only 14 B-17s in the 463rd returned to Italy. Six others were forced to land in Yugoslavia with badly wounded crew and mechanical problems.

On 15 April 1,142 heavy bombers dropped the 15th AF's largest bomb tonnage in a single 24 hour period. Some 830 B-17s and B-24s, escorted by 145 fighters, hit enemy gun positions, supply dumps, and troop concentrations, maintenance installations and military command centres along roads near Bologna. Another force of 312 B-17s and B-24s, escorted by 191 fighters, bombed railway bridges and an ammunition factory and stores. The next day, bad weather forced almost 700 B-17s and B-24s to abort the mission to the south-west of Bologna, but the strike went ahead on the 17th, when 751 heavies raided enemy positions closse to the city. Missions to the Bologna area continued on the 18th, and three days later Bologna was captured by ground forces. On the 21st, 240 B-17s and B-24s, with a P-51 escort, bombed marshalling yards in Germany and Austria. On 23 and 24 April, over 700 B-17s and B-24s bombed road and rail bridges over the Brenta and Adige rivers.

B-17G-55-DL 44-6606 of the 2nd Bomb Group landed at Miskolc airfield, Hungary, early in December 1944 after being damaged during a raid on Vienna. Miskolc was in an area liberated by Soviet and Romanian troops, and was the home base of the 8th Romanian Assault Group equipped with Henschel Hs 129 aircraft. The B-17 was repaired by Romanian groundcrews, who fixed an Hs 129 main wheel to the Fortress to act as a tail-wheel so it could take off again. Note the fuselage star applied aft of the right waist gun: this became necessary with the adoption of staggered waist guns on late B-17Gs. *(Hans-Heiri Stapfer)*

B-17s of the 15th Air Force attack the heavily-defended Schwechat oil refinery near Vienna, Austria amid thick flak. 15th AF machines were primarily used against German targets in the south and east of the Reich, which were generally difficult to reach for 8th AF aircraft based in Britain *(USAF)*

A B-17G of the 15th Air Force over its target takes a near miss from a German flak burst. *(USAF)*

On 25 April, 467 B-17s and B-24s bombed marshalling yards and rail networks in Austria. The next day, bad weather caused the force of 117 B-17s and 196 B-24s to abort their mission to targets in northern Italy. Bad weather kept them gounded for the next four days. Finally, on 29 April, the German forces in Italy surrendered, and the next day Hitler committed suicide in a bunker deep beneath the Reich's Chancellery. The war went on, however, and on 1 May, despite bad weather, 27 B-17s bombed marshalling yards at Salzburg. This was the final 15th AF bombing mission of the war.

The 15th Air Force had dropped 303,842 tons of bombs on Axis targets in a dozen countries, including major installations in eight capital cities. Some 148,955 heavy bomber sorties were flown. The contribution it made towards achieving the final victory in Europe was therefore of the highest importance.

B-17G-110-BO 43-39335, a VIP aircraft, taking off from Pilsen, Czechoslovakia, in 1946. *(Zdenek Hurt)*

3. Operational History 2: European Theatre 1942 to 1945

In 1942 American and RAF air leaders disagreed on the best method for employing strategic air forces against Germany. The British wanted the 8th Air Force to join RAF Bomber Command in its night bombing offensive, but the USAAF was determined to pursue its daylight precision bombing strategy. By early August 1942, General Ira C. Eaker, Commander of VIIIth Bomber Command in Britain, could call upon just three B-17 bomb groups, the 92nd, 97th and 301st, and only the 97th was considered operational. It had the honour of flying the first heavy bomber mission, when twelve B-17Es attacked Rouen on 17 August. Six other B-17s flew a diversion.

On 5 September the 301st Bomb Group flew its first mission when thirty-seven B-17s attacked the marshalling yards at Rouen. On 6 September the largest bombing mission so far took place when 36 B-17s of the 92nd and 97th Bomb Groups raided the Avions Potez factory at Meaulte while the 301st Bomb Group flew a diversionary raid on St Omer-Longuenesse airfield. Four squadrons of RAF Spitfire IXs escorted them but they failed to rendezvous with the bombers and about 50 enemy fighters intercepted the B-17s. Two B-17s were shot down and on two other B-17s three gunners were killed and five others wounded. Further missions followed, to the Rotterdam shipyards, to Cherbourg, to St Omer/Longuenesse airfield and to Meaulte.

On 9 October, 84 B-17s, including 24 from the 306th Bomb Group at Thurleigh, Bedfordshire, went to the huge steel and locomotive works at Lille. Each Fortress in the 306th Bomb Group had a crew of nine, a second waist gunner not being thought necessary as it was believed the gun on the inside of the formation would not need to be used! The 306th Bomb Group lost four B-17s to fighters and the 301st and 92nd Bomb Groups each lost one B-17 to fighters. Only sixty-nine bombers hit their primary targets and many of the bombs failed to explode. Many bombs fell outside the target area, killing a number of French civilians. Crews claimed forty-eight fighters destroyed, eighteen probably destroyed and four damaged. In fact the Germans lost only one fighter. At the time the figures did much to compensate for the largely inaccurate bombing.

European objectives

On 20 October 1942 Brigadier-General Asa N. Duncan, Chief of the Air Staff, issued a revised set of objectives to be carried out by VIIIth BC. In part it stated... 'Until further orders, every effort of the VIIIth Bomber Command will be directed to obtaining the maximum destruction of the submarine bases in the Bay of Biscay...' On 21 October 83 B-17s of the 97th, 301st, and 306th Bomb Groups were despatched to the U-boat pens at Keroman, about eleven miles from Lorient, while 17 B-17s of the 11th Combat Crew replacement Centre were to bomb Cherbourg. The two forces flew a long over-water flight in the Bay of Biscay to reduce the threat of *Luftwaffe* interception. However, thick cloud at their prescribed bombing altitude of 22,000 ft forced all except the fifteen Fortresses

B-17E 41-9129, which arrived at Prestwick, Scotland, on 24 July 1942 after repairs to mend a broken tail wheel, sustained in an accident at Bathurst, New Brunswick. This Fortress was assigned to the 359th Bomb Squadron, 303rd Bomb Group, forming the BOLERO weather flight, and went on to serve the 97th, 92nd and 305th Bomb Groups in the 8th Air Force before leaving the 1st Bomb Wing in 1943. *(USAF)*

of the 97th Bomb Group to return to England. The 97th Bomb Group bombed the target but just after re-crossing the French coast they were bounced by a swarm of yellow-nosed Fw 190s. Three B-17s were shot down and six were badly damaged. The 97th now had the highest losses of any group to date.

U-boat pens

On 7 November the 91st Bomb Group flew its first mission when 68 B-17s and B-24s went to the U-boat pens at Brest. Flak and fighters shot down one 306th Bomb Group B-17. Missions to the U-boat pens continued and losses mounted. On 9 November, three Fortresses in the 306th Bomb Group were lost at St Nazaire. On 18 November the 303rd Bomb Group flew its first mission and on 22 November Col Curtis E. LeMay's 305th Bomb Group flew its inaugural mission, part of a force of 68 B-17s and eight Liberators, to Lorient. Just 11 B-17s of the 301st Bomb Group formation managed to get their bombs away.

LeMay was determined to improve bombing accuracy and insisted his crews flew a straight course on the bomb run instead of zigzagging every ten seconds; a tactic which had been designed to spoil the aim of the German heavy anti-aircraft batteries.

He tried this formation for the first time on 23 November, when VIIIth BC went to St Nazaire again. Bad weather and mechanical problems forced several bombers to abort but LeMay's tactics worked and none in the 305th Bomb Group were lost. Other groups suffered however. The *Luftwaffe* had revised its tactics too, concentrating their attacks on the frontal area of a B-17 where the defensive firepower was weakest. In one pass JG2 knocked down four B-17s and a fifth victim staggered home to crash in England.

Small-scale raids continued against the U-boat pens and airfields in France and by the year-end, formations of 80 plus Fortresses became usual. Early in January 1943 Eaker abandoned individual bombing, which had been SOP (Standard Operating Procedure) in favour of Group bombing because 'lead crews' and 'staggered formations', developed by Col LeMay, proved more effective, although losses reached a new high. The growing strength of VIIIth BC increased pressure on the *Luftwaffe*, which now attacked in larger formations, and simultaneous attacks by fighters, rather than in trail, were used. Head-on attacks from twelve o'clock high became the norm.

On 27 January 1943, VIIIth BC attacked Germany for the first time. Sixty-four B-17s and Liberators started the mission, to the naval base at Wilhelmshaven on the northwest German coast. Having lost their way over the North Sea, 23 Liberators of the 44th and 93rd Bomb Groups wrongly identified Lemmer in Holland as Wilhelmshaven, dropping their bombs to the north of this harbour town. Fifty-five Fortresses reached Wilhelmshaven. The 306th Bomb Group led the mission with Col Frank A. Armstrong in the lead ship. Crews reported being attacked by about fifty fighters, including some twin-engined types. On 4 February enemy fighters claimed six B-17s shot down.

On 16 February 77 B-17s and 18 B-24s were dispatched to bomb the U-boat pens at St Nazaire. *Luftwaffe* fighter controllers waited until the escort fighters turned back before attacking. Six B-17s – two each from the 303rd, 305th and 306th Bomb Groups – were shot down. Gunners aboard the bombers claimed twenty enemy fighters shot down but in fact only one enemy fighter was written off. Thirty heavies returned battle damaged.

Fighter escort

During February and March 1943 aircraft factories, U-boat bases and marshalling yards were bombed. On 8 March P-47C Thunderbolts escorted the bombers for the first time. On 18 March seventy-three Fortresses and twenty-four Liberators, the highest number of bombers so far, attacked shipbuilding yards a few miles north of Bremen and on 22 March 84 Liberators and Fortresses attacked U-boat yards at Wilhelmshaven. With escorts and rising bomber strengths, missions were flown almost daily. On 31 March, 102 B-17s and B-24s were despatched to bomb the docks area at Rotterdam but four bomb groups were recalled because of strong winds and thick cloud. Two 303rd Bomb Group B-17s were lost in a mid-air collision. Thirty-three bombers hit the dock area but many heavies, blown off course by

Bombs dropped by the 305th Bomb Group hurtle down onto Vegesack on 18 March 1943. Flying at 24,000 feet, 76 per cent of the bombs exploded within 1,000 feet of the mean point of impact. *(USAF)*

Although B-17F-10-BO 41-24485 *Memphis Belle* of the 324th BS, 91st Bomb Group at Bassingbourn was not the first Fortress in the ETO to achieve 25 missions, it nonetheless served Hollywood and the Army Air Forces' publicity machine well. After the *Belle's* 25th mission, to Lorient, on 17 May 1943, she returned to the USA to begin a War Bond Tour. The movie which made the Fortress famous was first shown in April 1944. *(USAF)*

strong winds and bad visibility, missed their objectives completely and killed 326 Dutch civilians when their bombs exploded in the streets of Rotterdam.

During April attacks were made on industrial targets in France and Belgium. On 17 April, a record 111 bombers bombed the Focke Wulf factory at Bremen. Intense fighter attacks shot down fifteen Fortresses, and thirty-nine more suffered battle damage. During May attacks were made on U-boat pens, and industrial targets in France and Belgium. On 14 May the 8th AF made simultaneous attacks on four targets, but the day's missions cost twelve B-17s and B-24s. Sixty-seven fighters were claimed shot down.

Multi-pronged attacks

Through June 1943 the practice of making multiple attacks continued. Eaker was able to send the 1st and 4th Bomb Wings (BWs) on two-pronged attacks against north German targets at Emden, Kiel, Bremen, Wilhelmshaven and Cuxhaven on a single day. On 11 June the

B-17s set out to bomb Bremen but the target was covered with a solid layer of low cumulus clouds down to about 5,000 ft and they also partially covered Wilhelmshaven. Some 168 B-17s bombed Wilhelmshaven, while 30 bombed the secondary target at Cuxhaven. About 20 fighters attacked the low groups in the 4th BW but it was the 1st BW that suffered the greatest onslaught. They were attacked by fighters for over an hour. The 1st BW lost eight B-17s in all and the 4th BW, one Fortress. Sixty-two B-17s returned damaged. The American gunners claimed to have shot down 85 enemy fighters. In actual fact, only seven German fighters were destroyed or damaged and two pilots injured.

On 22 June the 8th made its first large-scale attack on the Ruhr, when 182 heavies bombed the chemical and synthetic rubber plant at Hüls. Three new B-17 groups – the 100th, 381st and 384th Bomb Groups – flew their first mission this day. Sixteen B-17s were shot down.

Despite continuing losses, the addition of more groups allowed Eaker to send a record 322 bombers to Hanover on 17 July. On 24 July

B-17F-35-DL 42-3190 of the 331st BS, 94th Bomb Group, which pilot Captain Kee Harrison crash-landed in a French wheat field with its bomb load intact after being shot up on the Paris-Le Bourget mission of 14 July 1943. Harrison and three others evaded the searching Germans and later returned safely to England. Four other members of the crew were made prisoners of war. *(via Zdenek Hurt)*

a week-long series of heavy bomber raids, later called 'Blitz Week', went ahead, beginning with a raid by 208 B-17s on Heroya and Trondheim in Norway. One B-17 failed to return and 64 were battle damaged. Next day, 25 July, 218 bombers attacked the ports of Kiel and Hamburg, devastated the night before in the great RAF Bomber Command fire raid.

Destruction of Hamburg

Cloud cover and the huge smoke pall from the still burning fires at Hamburg caused 59 B-17s of the 92nd, 305th and 306th Bomb Groups to abandon their strike altogether. The Hamburg force was intercepted as the leading B-17s approached the Elbe estuary and in running battles, fifteen B-17s – seven of them from the 384th Bomb Group – were shot down, and 67 returned battle damaged. The Kiel force lost four B-17s and 50 returned damaged, two of which crashed on landing.

On 26 July 92 heavies bombed rubber factories at Hanover and 54 others attacked shipbuilding yards at Hamburg. Twenty-four

aircraft were lost, mostly to enemy fighters. After a stand down on 27 July the 8th despatched just over 300 bombers in two forces to bomb German targets on the 28th, but bad weather interfered with the mission and only forty-nine bombed the Fieseler aircraft works at Kassel, with twenty-eight hitting the Fw 190 factory at Oschersleben. Twenty-two bombers were lost. On 29 July 168 B-17s of the 1st BW were despatched to bomb the U-boat yards at Kiel and 81 Fortresses of the 4th BW attacked the Heinkel aircraft factory at Warnemünde. The 4th BW lost four Fortresses and the 1st BW lost six B-17s. Four of the 1st Wing's losses came from the 306th Bomb Group. On 30 July 186 B-17s went to Kassel. Altogether, twelve bombers and six US fighters were lost. Next day, 31 July, VIIIth BC announced a three-day stand down from combat. In a week of sustained operations, about 100 aircraft and 90 combat crews had been lost.

On 12 August 243 heavy bombers were despatched unescorted to targets in the Ruhr. Some twenty-five bombers were shot down.

On 17 August 1943, 376 8th AF B-17s bombed the Schweinfurt ball bearing plant and the aircraft plants at Regensburg (pictured). Sixty B-17s were shot down, 36 at Schweinfurt and 24 at Regensburg. 27 B-17s were so badly damaged that they never flew again; and 60 B-17s of the 4th BW, which continued to North Africa, had to be left there for repair. *(USAF)*

Over 300 bombers were sent to attack targets in Holland and France on 15 August. This was part of the *Starkey* deception plan which was created to relieve some of the pressure on Russia and halt troop movements to Italy. The plan was intended to make the Germans believe that an invasion of the French coast was imminent. Strikes against enemy airfields in France and the Low Countries continued on 16 August, then early that evening base operation staff throughout eastern England waited for their orders for the morrow; the anniversary mission of the 8th AF.

Eaker and his planners had conceived a most ambitious and daring plan to attack, simultaneously, aircraft plants at Schweinfurt and Regensburg. The *Luftwaffe's* operational fighter strength on the western front was showing a significant increase and Regensburg was the second largest aircraft plant of its kind in Europe, the largest being at Wiener-Neustadt near Vienna. The original plan to bomb all three plants on one day, 7 August, had been disrupted by bad weather so the plan had been modified to bomb each target when the opportunity arose. On 13 August Wiener-Neustadt was bombed by B-24s of VIIIth BC and on 14 August by B-24s of the 9th AF, both forces flying from North Africa.

Targeting fighter factories

Production at Regensburg was estimated at 200 Bf 109s a month, or approximately 25 to 30 per cent of Germany's single engine aircraft production at that time. It was estimated that the total destruction of the Regensburg plant would enforce a nine-month delay in production. It was hoped that immediate results would be felt in operational strength, within one and a half to two months. The double strike was a bold move as hitherto, the campaign against the German aircraft industry had been waged within reasonable striking distance from the British mainland.

Brigadier-General Robert Williams commanded the 1st BW, and he would lead his force to Schweinfurt while Col Curtis E. LeMay would lead the 4th BW to Regensburg. To minimise attacks from enemy fighters it was decided that LeMay's B-17s would fly on to North Africa after the attack. The 1st BW, meanwhile, would fly a parallel course to Schweinfurt to further confuse the enemy defences and return to England after the raid. Few of the 1st BW Fortresses were equipped with 'Tokyo tanks', and without them could not make the 725 mile trip to North Africa.

Despite the careful route planning, Eaker and his subordinates knew the B-17 crews would have a running fight on their hands. However, they hoped that the fighter escort would keep losses down. Four P-47 groups were scheduled to escort the Regensburg force, but only one group rendezvoused with the bombers as scheduled. The overburdened Thunderbolts could not possibly hope to protect all seven groups in the 4th BW. The long, straggling formation stretched for fifteen miles and presented the fighter pilots with an awesome responsibility. Fortresses in the rear of the formation were left without protection.

Focke-Wulf Fw 190s of JG1 and JG26 began their attacks 30 km east of Brussels. In the ninety minutes preceding the bomb run, seventeen Fortresses were shot down. The bombing was extremely accurate, which might well have had something to do with the presence of Col LeMay, exponent of high-level bombing techniques, in the first wave. Six main workshops were hit, five being severely damaged. A hangar was partially destroyed and storerooms and administrative buildings wrecked. Thirty-seven Bf 109s at dispersal were at least damaged if not wrecked and all production at the plant came to an abrupt halt. Although unknown at the time, by way of a bonus, the bombing had destroyed the fuselage jigs for a secret jet fighter, the Me 262.

The surviving 128 B-17s, some flying on three engines and many trailing smoke, were attacked by a few fighters on the way to the Alps. LeMay circled his formation over Lake Garda near Verona to give the cripples a chance to rejoin the wing. Red lights were showing on all four fuel tanks in every ship and it was a

B-17-75-BO 42-29854 *Ole Flak Sack* of the 532nd Bomb Squadron, 381st Bomb Group, seen after belly-landing at Ridgewell in the summer of 1943. *Ole Flak Sack* was repaired, but was lost with Lt Arthur M. Sample's crew on the 8 October 1943 mission to Bremen, Germany. *(via Mike Bailey)*

ragged collection of survivors, which landed at intervals up to 50 miles along the North African coast. Altogether, the 4th BW had lost 24 bombers, with the 100th Bomb Group's nine losses the highest loss of all.

'Swinish Schweinfurt'

The 4th BW encountered so many fighters en route because thick inland mists had delayed the 1st BW by three and a half hours after the 4th BW had taken off. This had effectively prevented a two-pronged assault which might have split the opposing fighter force. The delay gave the *Luftwaffe* time to refuel, re-arm, and re-deploy to forward bases in Holland after dealing with the Regensburg force, and the 1st BW was hit hard. Attacks on the Schweinfurt force cost 36 Fortresses. The worst hit B-17 groups were the 381st and 91st, which lost eleven and ten B-17s respectively. Twenty-seven

B-17s in the 1st BW were so badly damaged that they never flew again. Altogether, sixty B-17s were Missing in Action (MIA) – almost three times as high as the previous highest casualty rate, on 13 June, when 26 bombers were lost. A further sixty Fortresses had to be left in North Africa pending repairs, so in the final analysis, 147 Fortresses had been lost to all causes on 17 August. Claims by the B-17 gunners and escorting fighters of 288 German fighters destroyed were wildly inflated as usual – the *Luftwaffe* actually lost 27 fighters on 17 August! The almost non-existent maintenance facilities in North Africa ruled out any further shuttle missions.

For a time the Schweinfurt losses had a mighty effect on the B-17 groups but the P-47D, with the installation of a 108-gallon belly tank, could now escort the bombers further. *Carpet* radar jammers on B-17s also helped confuse

B-17Fs of the 535th Bomb Squadron, 381st Bomb Group in formation. The nearest aircraft is B-17F-80-BO 42-29941/R. *TS Tarfu* – 'TS' standing in bowdlerised form for 'Tough Stuff' but which more commonly had a more lavatorial meaning. *TS Tarfu* failed to return from a mission with Capt Edwin R. Manchester Jr's crew on 8 October 1943. Six men were killed in action (KIA) and four were made Prisoners of War. *(USAF)*

On 9 October 1943, 8th AF bombers were despatched to targets in East Prussia and Poland on the longest mission up to that time. Some 115 aircraft were despatched to the Arado aircraft component plant at Anklam near Peenemünde as a diversion for 263 4th BW bombers attacking the Polish port of Gdynia and the Focke-Wulf plant at Marienberg. Anti-aircraft defences at Marienberg were limited, so the force bombed from under-13,000 ft, which resulted in 60% of the bombs dropped by the 96 Fortresses exploding within 1,000 ft of the MPI; 83% fell within 2,000 ft. Before the raid, the plant was responsible for almost half the *Luftwaffe's* Fw 190 production. Eaker called it 'A classic example of precision bombing'. *(C. Roy Babb via Abe Dolim)*

enemy radar. However, despite round-the-clock bombing of aircraft production plants, the *Luftwaffe* had first-line strength of some 1,525 single and twin-engined fighters for the defence of the western approaches to Germany. This was some 400 more than before the issue of the *Pointblank* directive, although only about a third of this force was ready for use, the rest being reserves or temporarily unserviceable.

Schweinfurt again

Early in October, the decision was taken to attack the ball bearing plant at Schweinfurt for the second time in three months, in what was hoped would be a single, decisive blow against the German aircraft industry, stemming the flow of fighters to the *Luftwaffe*. The mission went ahead on 14 October but the hoped for force of 420 Fortresses and Liberators making a three-pronged attack was thwarted by bad weather and aborts. Ultimately, 320 B-17s were

despatched. Schweinfurt soaked up 482.8 tons of high explosives and incendiaries, but losses were again high. The 1st BD lost 45 B-17s and the 3rd BD lost 15 Fortresses. In total, sixty Fortresses and 600 men were missing. Five B-17s had crashed in England as a result of their battle-damaged condition and twelve more were destroyed in crash landings or so badly damaged that they had to be written off. Of the returning bombers, 121 required repairs. Claims of 186 enemy fighters shot down were submitted (the actual figure was about 35). Only 88 out of the 1,222 bombs dropped actually fell on the plants and production at the Kugelfischer plant, largest of the five plants, was interrupted for only six weeks.

The losses and a spell of bad weather restricted VIIIth BC to just two more missions in October. After Schweinfurt, desperate attempts were made to improve the range of the few Thunderbolts in the ETO. Still the long-

B-17G-1-BO 42-31118 *Lady Millicent II* of the 338th Bomb Squadron, 96th Bomb Group, which bellied in at Snetterton Heath on 8 January 1944 and was salvaged five days later. *(Joseph Minton via Larry Goldstein)*

range missions continued. On 3 November, 566 B-17s and B-24s were despatched to Wilhelmshaven. Seven bombers were lost in a fierce air battle. Two days later 323 B-17s bombed Gelsenkirchen and 104 B-24s hit Münster. Eight B-17s were lost.

Bad weather pause

For the first two weeks of November 1943 England was blanketed by thick woolly fog and airfields were lashed with intermittent showers and high winds. When the bad weather front finally lifted on 16 November VIIIth BC struck at targets in Norway. The 1st BD attacked the molybdenum mines at Knaben and the 3rd BD attacked a generating plant at Vermark in the Rjukan Valley. Both targets were connected with the German heavy water experiments, which were designed to give the Nazis the atomic bomb.

On 31 December, when VIIIth BC celebrated its second year in England with a series of all-out raids on airfields in France, twenty-five bomber crews and four fighter pilots would not return to their bases to celebrate the New Year.

Early in 1944 the *Luftwaffe* was still a force to be reckoned with, but the long-range P-51 Mustang was now in service. This superb fighter could accompany the bombers to their targets and back again. Gen Carl 'Tooey' Spaatz and his subordinate commanders, Maj-Gen Jimmy Doolittle (8th AF) and Maj-Gen Nathan F. Twining (15th AF), planned to make a series of co-ordinated raids on the German aircraft industry, supported by RAF night bombing, at the earliest possible date. Good weather finally permitted Operation *Argument*, which soon became known as 'Big Week', to take place during the week 20–25 February.

On 20 February 1,028 B-17s and B-24s and 832 fighters in the 8th AF attacked twelve aircraft plants in Germany for the loss of 25 bombers and four fighters. Three Medals of Honor (two posthumously) were awarded to B-17 crewmen. The next day, 21 February, 924 bombers and 679 fighters bombed aircraft

B-17Gs of the 360th Bomb Squadron, 303rd Bomb Group, 8th Air Force en route to bomb transportation targets. By this time, the bombers were covered all the way by their fighter shield above. *(USAF)*

factories at Brunswick and other targets. The 8th lost 19 bombers and five fighters but 60 German fighters were claimed shot down. On 22 February the 8th bombed targets in Germany and Holland with the loss of 41 bombers. On 23 February bad weather kept the heavies on the ground. The next day, 24 February, 238 Fortresses attacked Schweinfurt, losing eleven, while 295 B-17s struck at targets on the Baltic coast with the loss of five Fortresses. On 25 February the USSTAF brought the curtain down on 'Big Week' when 1,300 8th and 15th AF bombers and 1,000 fighters were despatched to aircraft plants, ball bearing works and components factories throughout the Reich. The 1st BD caused heavy

damage to the Messerschmitt plants at Augsburg and the ball bearing plants at Stuttgart were also bombed. Although the 8th lost 31 bombers, the 3rd BD and 176 bombers of the 15th AF caused considerable damage. Output at both Augsburg and Regensburg was severely reduced for four months following the raids. In all, 'Big Week' cost 226 bombers.

Target Berlin

Less than a week later, the 8th AF bombed 'Big-B' – Berlin – for the first time. An attack on 3 March was aborted because of bad weather, and the next day severe weather en route prevented 502 Fortresses and 770 fighters to continue to the target. However, 30 B-17s in the

B-17F-75-BO 42-29837 *Lady Luck* and B-17F-15-BO 41-24490 *Jack the Ripper* of the 324th Bomb Squadron, 91st Bomb Group over France en route to their target at Tours on 5 January 1944. On 22 February 1944 *Jack the Ripper* became the first of the original B-17Fs assigned to the 91st Bomb Group to be shot down, being lost on a raid on Bunde near Osnabruck. *Lady Luck* joined AFSC on 6 April 1944 and then transferred to the MTO where it operated in Italy before returning to the US in June 1944. It was scrapped in July 1945. *(USAF)*

95th and 100th Bomb Groups defied the elements to drop the first American bombs on the German capital. On 6 March the 8th despatched 730 B-17s and B-24s and 801 P-38, P-47 and P-51 escort fighters to targets in the suburbs of Berlin. US fighters claimed 81 enemy fighters shot down and the bomber gunners claimed 97 destroyed. The *Luftwaffe* actually lost 64 fighters destroyed and two damaged beyond repair, much lower than had been claimed but still a considerable loss. The 8th lost a record 69 bombers and 11 fighters, while 102 bombers were seriously damaged.

Battle of Berlin

The Bomb Groups stood down on the 7th, before resuming the daylight offensive against Berlin again on 8 March. A total of 623 bombers were sent to bomb the VKF ball-bearing plant at Erkner in the suburbs of Berlin, escorted by 891 fighters. Of the force, 539 heavies got their bombs away over the German capital, which once again was heavily defended. The leading 3rd BD lost 23 Fortresses, 16 of them from the leading 45th CBW, which came under mass enemy fighter attack. The 1st BD lost five B-17s to fighter attacks and flak. Nine B-24s and 18 fighters were lost. The heavy bomber gunners claimed to have shot down 63 enemy fighters, while the escorts claimed a further 79 for the loss of 18 of their own. In fact, only 27 *Luftwaffe* fighters were lost on 8 March.

Despite the continued high losses, the 8th AF attacked Berlin again on the 9th. A total of 361 B-17s bombed Big B while 165 Liberators visited Hanover, Brunswick and Nienburg. For once the enemy fighters were noticeable by their absence – weather conditions had kept the *Jagdwaffe* on the ground. The 800 escort fighters returned without claiming any enemy fighters. Nevertheless, the B-17 groups lost six of their number over Berlin while the B-24 groups lost two, all to flak.

Smaller scale raids on targets in France and Germany followed. On 13 April 1944, overall command of the Combined Bomber Offensive and the 8th AF was temporarily transferred to Gen Dwight D. Eisenhower, newly appointed

On 11 April 1944, General Dwight D. Eisenhower christened B-17G-40-BO 42-97061 *General Ike*, at Bassingbourn, using a bottle of Mississippi River water. Major James McPartlin, 401st Bomb Squadron, was responsible for the naming of the Fortress, which flew its first operational mission on 13 April 1944, two days after it was christened. On 29 May 1944 *General Ike* led the 91st Bomb Group to Poznan, Poland, on the longest daylight raid of the war up to that time. *Ike* completed 75 combat missions, returning to the USA in June 1945 and being broken up for scrap. *(via Tom Cushing)*

Supreme Allied Commander. Missions in April and May alternated between targets in Germany and hitting pre-invasion targets in France, Luxembourg and Belgium. Operation *Cover* called for raids on coastal defences, mainly in the Pas de Calais, to deceive the Germans as to the area to be invaded by the Allied armies massing in Britain. On 6 June 1944, D-Day, a total of 2,362 bomber sorties, involving 1,729 B-17s and B-24s, was flown, dropping 3,596 tons of bombs.

US ground crews worked through the night of 6 June and all day on the 7th so that two missions could be flown. On 8 June 1,135 bombers were despatched to communication targets in France. Bad weather prevented 400 heavies from bombing, and the next day postponed any bomber strikes at all. It also severely curtailed operations on 10 June. Of the 873 bombers airborne, over 200 were forced to abort because of cloud conditions. Some 589 bombers, including 31 Pathfinders, attacked eight airfields in France and nine coastal installations in the Pas de Calais. On 11 and 12 June bad weather ruled out targets in Germany and the 8th despatched its bombers to France

B-17G-80-BO 43-38183 *Ice Cold Katie* of the 332nd Bomb Squadron, 94th Bomb Group at Bury St Edmunds (Rougham) in July 1944. This Fortress was later re-assigned to the 379th Bomb Group at Kimbolton. *(USAF)*

again. Tactical targets in France continued to be attacked until 15 June, when 1,225 bombers attacked an oil refinery at Misburg, and the 1st BD struck mostly at airfield targets in northern France, to prevent their being used by the *Luftwaffe* to launch attacks on the Normandy beach head.

Close Support in the Ardennes

From June onwards, mass formations of over a thousand bombers were not uncommon as the 8th carried out post invasion support missions and bombed targets throughout the Reich. It seemed that the war would be over by Christmas. However, on 16 December 1944, using appalling weather conditions to his advantage, Field Marshal Karl von Rundstedt and his panzer formations attacked American

positions in the forests of the Ardennes on the French-Belgian border. Supported by an estimated 1,400 German fighters, they opened up a salient or bulge in the Allied lines. The Allied air forces were grounded by fog and it was not until 23 December that they could offer bomber support in the 'Battle of the Bulge.' On Christmas Eve a record 2,034 8th AF bombers and 500 RAF and 9th AF bombers, took part in the largest single strike flown by the Allied Air Forces in WW2, against German airfields and communications leading to the Ardennes. One of the B-17 losses this day carried Brig-Gen Fred Castle, commander of the 4th BW, who was leading the 3rd BD. Brig-Gen Castle was posthumously awarded the Medal of Honor; the highest ranking officer in the 8th AF to receive the award. Overall, the Christmas Eve

B-17Gs of the 401st Bomb Squadron, 91st BG in formation. B-17G-65-BO 43-37624 *Victory Queen* nearest the camera, was lost on the mission to Leipzig on 20 July 1944 after being renamed *Liberty Queen*. (USAF)

raids were effective and severely hampered von Rundstedt's lines of communication.

Battle of attrition

German production of fighter aircraft actually increased through 1944 into 1945. It had peaked in September 1944, when an astonishing 1,874 Bf 109s and 1,002 Fw 190s were completed. However, in that same month, an average of three German fighters – and two pilots KIA – were lost for every B-17 or B-24 shot down. In spite of vastly inflated kill claims, the USSTAF was clearly winning the battle of attrition.

January 1945 marked the Eighth's third year of operations and it seemed as if the end of the war was in sight. Finally, the German advance in the Ardennes came to a halt and ultimately petered out. Hitler's last chance now lay in his so-called 'wonder weapons' – the V1 and V2. Missions were flown to tactical targets throughout the remaining days of January, but when the weather intervened, the 8th Air Force mounted shallow penetration raids on V-weapon targets in France under the codename *No-ball*. The 8th also attempted several tactical missions but the weather was so bad morale sagged as mission after mission was scrubbed, often just after take-off.

By 3 February 1945 the Red Army was only 35 miles from Berlin and the capital was jammed with refugees fleeing from the advancing Russians. Accompanied by 900 fighters, 1,200 B-17s and B-24s dropped 2,267 tons of bombs on the centre of Berlin, killing an estimated 25,000 inhabitants and destroying 360 industrial firms, heavily damaging another

B-17G-45-BO 42-97271 *Boss Lady* of the 545th Bomb Squadron, 384th Bomb Group flies in formation with other 'Gs' en route to their target at Frankfurt. *Boss Lady* went MIA on the mission to Hamburg on 30 March 1945. One of Lt Hicks' crew was KIA, and eight were taken prisoner. *(USAF)*

170. But the *Luftwaffe* was still far from dead. The 8th Air Force lost 21 bombers shot down; another six crash-landed inside the Russian lines. Of the bombers that returned, 93 had suffered varying forms of major flak damage. On 9 February the heavies returned to the oil refineries in the ever-diminishing Reich.

Final destruction

On 22 February Operation *Clarion*, the destruction of what remained of the German communications network, was launched. More than 6,000 aircraft struck at transportation targets in western Germany and northern Holland. Only seven bombers, including two B-17s, were lost. By March 1945 the systematic destruction of German oil production plants had virtually driven the *Luftwaffe* from German skies, although Me 262 jet fighters could still be expected to put in rare attacks.

On 23/24 March, under a 66 mile long smoke screen and aided by 1,749 bombers of the 8th AF, Field Marshal Bernard Montgomery's 21st Army Group crossed the northern Rhine, while further south simultaneous crossings were made by General Patton's Third Army.

Beginning on 5 April, the weather over the continent improved dramatically and the B-17s were despatched to U-boat pens on the Baltic coast. Everywhere the Allies were victorious. In Germany, last ditch attempts were made by the *Luftwaffe* to try to stem the tide and even the deliberate ramming of American bombers by converted Bf 109 and Fw 190 fighters, called *Rammjäger*, was tried. They could not prevent the final outcome however.

During the week following 18 April, 8th Air Force missions were briefed and scrubbed almost simultaneously as Allied ground forces overran objective after objective. The end came on 25 April 1945 when 306 B-17s of the 1st AD bombed the Skoda armaments factory at Pilsen in Czechoslovakia.

During the first week of May the German armies surrendered and in May Fortress crews flew mercy missions, called *Chowhound*, to starving Dutch civilians in Holland. VE (Victory in Europe) Day took place on 8 May.

4. Accomplishments: The Fortress Men

The Fortress was tough. There are many tales of B-17s that returned to their bases in Europe, the Mediterranean, and the Pacific, 'on a wing and a prayer'. Their crews too earned undying fame for their bravery and fortitude, bombing targets against all odds, and stories of crews returning in often shattered and bullet-riddled Forts are legendary. And the legends started early: the first tales of American bravery in the Pacific reached the attention of a shell-shocked public, desperate for stories to boost morale, just a few weeks after Pearl Harbor.

On 10 December 1941 five B-17Cs of the 19th Bomb Group led by Major 'Rosie' O'Donnell, 14th Squadron CO, carried out the first American bombing raid of the war when they attacked a Japanese convoy landing troops and equipment at Vigan and at Aparri in northern Luzon in the Philippines. Armed with mixed bomb loads, the B-17s valiantly tried to destroy enemy shipping but no targets were hit, and one of the B-17s was ditched. Another was attacked by Zeros and was badly hit and a third landed damaged with one engine out.

A B-17C piloted by Capt John P. Kelly Jr, carried only three 600-lb bombs. Kelly ignored the Japanese landing operations under way at Vigan, and carried on to Aparri in search of an enemy aircraft carrier, which had been reported. Finding no sign of the carrier, Kelly returned to Vigan and attacked the heavy cruiser *Ashigara* from 22,000 ft. One of the three bombs hit the aft gun turret, and started a fire. A group of Mitsubishi A6M Zeros gave chase,

and about fifty miles from Clark Field they caught up with the Fortress. Successive attacks destroyed parts of the aircraft, which then caught fire in the bomb-bay area. One waist gunner was decapitated by a burst of machine-gun fire, and another crewman was wounded. Kelly bravely battled to keep the B-17C straight and level while his co-pilot and four other crew evacuated the stricken aircraft. Despite being fired on by the circling Zeros, they all landed safely at Clark, but the Fortress finally exploded before Kelly could escape. Kelly was posthumously awarded the DSC and was later recommended for the Medal of Honor, for 'sinking' the Japanese battleship *Haruna*.

Sacrifice in the Pacific

The Japanese successfully established a bridgehead at Legaspi on southern Luzon and further raids by B-17Cs from Del Monte to attempt to dislodge them were small and ineffective. In one particularly brave attempt, 1/Lt Hewitt T. 'Shorty' Wheless dropped all eight 600-pounders on shipping while opposed by eighteen Zeros. During their attacks the Japanese fighters killed the belly gunner, and badly wounded three of Wheless' crew. Wheless somehow kept the stricken B-17 airborne. Badly shot up – the plane had more than 1,200 bullet holes – and losing fuel, the B-17 crash-landed at a small strip at Cagayan, twenty miles northwest of Del Monte. The B-17 stood on its nose before falling back on its tail but the wounded scrambled out safely. Wheless was awarded the DSC for the action.

Another 19th Bomb Group pilot, Capt Harl Pease Jr, was posthumously awarded the Medal of Honor for his actions during 6–7 Aug 1942. On 6 August Pease was forced to return to Mareeba from a reconnaissance mission over New Britain when one engine of his B-17 failed. Pease was anxious to take part in the big raid planned for the 7th against Vunakanau airfield in the Bismarck Archipelago, so he and his crew worked for hours on 41-2429, a replacement plane. They finally arrived at Port Moresby after midnight.

On the morning of 7 August, sixteen B-17s of the 19th Bomb Group, led by Lt Col Richard H. Carmichael, took off for Vunakanau airfield where 150 bombers threatened the US Marine Corps' landings on Guadalcanal. One B-17 crashed on take-off and two aborted with mechanical malfunctions, but by skilful flying, Pease maintained his position in the formation, despite a still troublesome engine, and made it to the target. At this point the bad engine gave out. Pease feathered it and dropped his bombs on the target, but the Japanese fighters soon singled out the lame duck. In the air battle that continued after the bombers left the target, Pease's B-17 was hit in the bomb-bay tank, which burst into flames, and the bomber fell behind the formation and was lost. There were reports of two parachutes being seen, and some years later it emerged that Pease and Sgt Czehowski, a gunner, did bale out and were taken prisoner. On 8 October 1942 Pease, Czehowski and four other prisoners were executed by the Japanese at Rabaul.

One against fifteen

Weeks earlier, on the morning of 16 June 1942, Capt Jay Zeamer Jr, aged 25, a pilot in the 65th Bomb Squadron, 43rd Bomb Group, set off on an aerial mapping sortie over the Solomon Islands. It was his 47th combat mission. The night before, the crew, all of whom were volunteers, was told to include a reconnaissance over Buka passage, as 400 enemy planes had just landed there. Zeamer arrived at the mapping site before the sun had risen high enough to take photos, and proceeded to Buka first. With just forty-five seconds of the mapping mission remaining, *Lucy*, their B-17E, was attacked by more than fifteen fighters. Although mortally wounded, 2/Lt Joseph Sarnoski, bombardier, remained at his nose-guns and fired at the enemy attackers until he died at his post. Though seriously wounded by shrapnel in his legs and both arms, Zeamer manoeuvred *Lucy* for forty minutes during the combat until the enemy broke off their action, then directed the flight to a base more than 500 miles away. When Zeamer passed out from loss of blood, Sgt John Able, the top turret gunner, took over, since co-pilot Lt John Britten was also injured. Zeamer, who was barely conscious, put *Lucy* down at Dobodura on New Guinea, with one crewman dead and five wounded. Zeamer hovered on the edge of death for three days, and spent fifteen months in more than a dozen hospitals. He and Sarnoski were both awarded the Medal of Honor. Radio operator Bill Vaughan, who while severely wounded managed to pick up a distant radio signal that allowed them an approximate heading to get them home – the navigator had been critically wounded – was awarded the DSC.

Medal of Honor over Ploesti

It was for actions like these that the Fortress is remembered, not only in the early battles in the Pacific but in Europe and the Mediterranean too. On 23 June 1944, 139 bombers of the 15th Air Force in Italy arrived over the oil refineries at Ploesti and dropped over 280 tons of bombs on the target. One of the six aircraft lost on this day was B-17F-35-VE 42-5951 *Opissonya*, part of the 341st BS, 97th BG. Its bombardier, 2/Lt David R. Kingsley, who was on his twentieth mission, was posthumously awarded the Medal of Honor for his heroism.

The Fortress was badly damaged by flak on the bomb run but Kingsley was still able to drop his bombs successfully. Three Bf 109s then attacked, damaging *Opissonya* still further. Sgt Michael J. Sullivan, the tail gunner, was badly wounded in the upper arm. Kingsley administered first aid. Meanwhile eight more 109s attacked and the ball gunner was hit by 20-mm shell fragments and badly wounded. When the pilot gave the order to bale out, Kingsley immediately assisted the wounded gunners – he even gave Sullivan his own parachute harness when the gunner's could not

Hollywood movie star and 8th AF gunnery training officer Clark Gable poses with Lt George Birdsong's crew of B-17F-30-BO 42-5077 *Delta Rebel No 2*, of the 323rd Bomb Squadron, 91st Bomb Group, at Bassingbourn. This B-17, then flown by 2/Lt Robert W. Thompson's crew, failed to return on 12 August 1943. *(via Tom Cushing)*

be found. He helped the wounded men bale out of the doomed B-17, and stayed with the aircraft, which was on automatic pilot, until it crashed and burned. His body was later discovered in the wreckage.

Twelve Medals of Honor, seven of them posthumous awards, were awarded to 8th Air Force personnel for 'conspicuous gallantry and intrepidity above and beyond the call of duty' while flying from England in WW2. The first MoH was posthumously awarded to 1/Lt Jack Mathis of the 303rd Bomb Group, for his actions on 18 March 1943. Capt Harold Stouse

in the 359th Bomb Squadron brought the badly damaged *Duchess* back to Molesworth with the dead body of lead bombardier Jack Mathis in the nose. Though mortally wounded over the target, Mathis, who was doing the aiming for all the other B-17s in the squadron, got his bombs away, and enabled the rest of the squadron to do likewise.

S/Sgt Maynard 'Snuffy' Smith became the first enlisted man in the 8th Air Force to receive the Medal of Honor, for his actions in an attack on St Nazaire on May Day 1943. The 306th Bomb Group lost six B-17Fs, and Lt Lewis P.

S/Sgt Maynard 'Snuffy' Smith, 306th Bomb Group, May 1943. Smith was the 8th AF's first enlisted Medal of Honor winner. *(USAF)*

Johnson Jr's aircraft was hit several times, catching fire in the radio compartment and in the tail area. Smith, the ball turret gunner, on his first mission, hand-cranked his turret to get it back into the aircraft; however, when he climbed out he discovered that the waist gunners and the radio operator had baled out. He, nevertheless, remained in the aircraft and fought the fire with a hand extinguisher. Moreover the Fortress did not show any signs of leaving formation, so he assumed the pilots

were still aboard and went to treat the badly wounded tail gunner. Then he jettisoned the oxygen bottles and ammunition in the radio compartment, and manned the waist guns during an attack by enemy fighters, only stopping to dampen down the fires and treat the tail gunner. Johnson put the B-17F down at Predannack near Land's End after Smith had thrown out all expendable equipment.

On 20 February 1944, during 'Big Week', three Medals of Honor were awarded to B-17 crewmen, the only instance in 8th AF's history when more than one was issued on one day. 1/Lt William R. Lawley Jr, 23 years old, was a pilot in the 364th Bomb Squadron, 305th Bomb Group, which raided Brunswick. He was decorated for getting his badly crippled B-17 and his crew back to England, after suffering serious injuries at the target. In the 351st Bomb Group, Sgt Archie Mathies, ball turret gunner, and 2/Lt Walter E. Truemper, navigator, valiantly brought *Mizpah* back to Polebrook after their pilots were killed, but died attempting to land.

The struggle of the *Bertie Lee*

Lieutenant Edward S. Michael of the 364th Bomb Squadron, 305th Bomb Group, brought *Bertie Lee* home to England after it had been devastated by cannon fire near Brunswick and had plummeted into a 3,000 ft dive, its bomb-bay on fire. Michael was seriously wounded in the thigh and his instruments were shattered. He and his co-pilot Lt Franklin Westberg finally got the B-17 out of the spin and Michael ordered the crew to bale out. Sgt John Leiber, bombardier, had lost an arm and was pushed out with a 'chute by the engineer, Sgt Jewell Phillips. Phillips' parachute, meanwhile, had been shredded by shrapnel. He and the two pilots were too busy to decide who would use whose 'chute. 'If we can't all jump together,' Michael screamed at Westberg, 'then we'll all go down together!'

Michael and Westberg however, managed to reach RAF Waltham near Grimsby. Fighting off unconsciousness, Michael performed a perfect belly landing, even though the undercarriage and flaps had been put out of operation, the ball turret was stuck in the lowered position with its guns pointing downwards, the

B-17F *Big Bust* of the 379th Bomb Group seen here at Kimbolton early in 1943. (*USAF*)

B-17F-70-DL 42-3494 of the 92nd Bomb Group which crashed at Podington on 6 September 1943 returning from a raid on Stuttgart. On that day the 92nd Bomb Group lost three crews; one more ditched in the Channel and four made forced landings at other bases in England. 42-3494 and 2/Lt Raymond J. Mancu's crew did not return from the Kiel raid on 4 January 1944, lost to flak at the target. (*Joseph Minton via Larry Goldstein*)

airspeed indicator was dead, and the bomb-bay doors were jammed fully open. Michael was hospitalized for seven weeks. Later, Michael was awarded the Medal of Honor, the second member of the 364th Bomb Squadron to receive America's highest gallantry award.

Damaged over the 'Big B'

The Fortress earned tributes from aircrews for the punishment it could take. William C. Stewart, 92nd BG, recalls a raid on Berlin.

'As we proceeded over 'Big B' the ship lurched and twisted from impacts, but we kept going and finally unloaded our bombs onto a railroad station in the heart of the city. Once away from the city, the flight back was uneventful. When I turned my turret to about two o'clock, I could see strips of aluminium skin peeled back from the right wing. When we landed we found we had taken what must have been a 105-mm anti-aircraft shell through the wing. It left a hole that a man could put his head through near the inboard end of the 'Tokyo gas tank'. There were hundreds of smaller holes throughout the main body, wings and tail sections but no one was hit. I had learned what accurate and intense flak was.'

Harry W. Love, bombardier, 568th Bomb Squadron, 390th Bomb Group, recalls:

'Over the Ruhr Valley we began to pick up massive concentrations of flak fire. A massive vibration developed and fumes and smoke filled the plane. The pilot, without hesitation, pulled out of formation, and attempted to put out the flames within that particular engine by sideslipping the plane. The entire right wing was oscillating up and down some 20–30 degrees. On seeing this, I assisted the navigator in putting on his chest pack. As bombardier, I always wore my backpack throughout the entire mission. Looking out at the wing again (which was oscillating even more), it was obvious to me that the wing could not stay on much longer. At this point, firmly believing the alarm bell and intercom were no longer operating, I directed the engineer to bail out. The engineer would have to bail out of the plane first, the navigator second and then myself. I ordered the engineer to bail out. He hesitated. I then began to physically push the navigator in that direction stating, "We have to go, the wing is coming off." The navigator looked at me with quite an acceptable (and understandable) look of doubt, and shook his head. At that instant, the wing came off.'

Richard Wynn, a navigator in the 'Bloody Hundredth' Bomb Group, recalls the 12 May 1944 raid on Brux, Czechoslovakia.

Lt Thomas H Gunn, 323rd BS, 91st Bomb Group, flew B-17G-20-DL 42-37938 *Betty Lou's Buggy* safely back to Bassingbourn on 19 April 1944 after sustaining heavy damage over the FW assembly plant at Eschwege near Kassel. Fighters holed the left wing fuel tank, put a turbo out, damaged an engine, knocked out the elevators and left aileron and exploded shells in the nose, cockpit, bomb-bay and fin, which severed the rudder controls. Gunn could only control direction by using the engines and had to apply full right aileron to keep the B-17 level. Only the co-pilot and navigator were wounded. *Betty Lou's Buggy* was repaired and survived the war, only to be cut up for scrap at Kingman, AZ, in December 1945. *(USAF)*

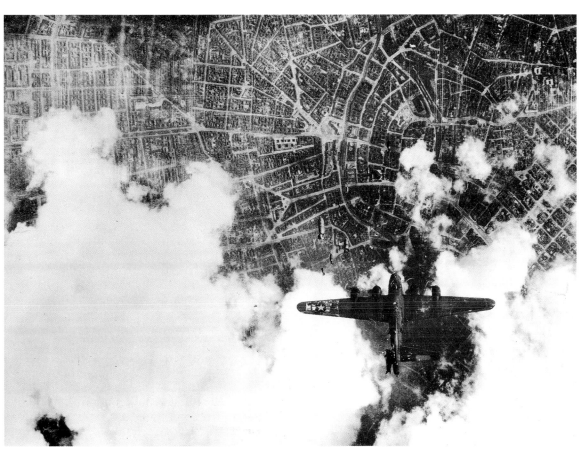

A stick of 1,000 lb bombs dropped from Lt John Winslett's B-17G-30-VE 42-97791 *Trudy*, 332nd BS, 94th Bomb Group, over Berlin, 19 May 1944, knocked off the left horizontal stabilizer of Lt Marion Ulysses Reid's B-17G-20-BO 42-31540 *Miss Donna Mae*, 331st BS, below. Reid's aircraft went into an uncontrollable spin. At 13,000 ft the wing broke off and the B-17 spun crazily to the ground. There were no survivors. *(USAF via Abe Dolim)*

'Our ship was the famed *Rosie's Riveters*, made famous, not by us, but by Robert Rosenthal, a legendary figure with the 100th. We inherited the ship from him when his crew finished its first tour. The airplane was really either the second or third to be known as *Rosie's Riveters*. We encountered light flak over the target but observed some rocket-propelled anti-aircraft fire. We were hit in No. 3 and lost oil pressure before the prop could be feathered. Cockpit procedure was complicated by the fact that a faulty flare from the lead ship had deposited a heavy film of "guck" over the pilot's half of the windshield putting the co-pilot in control of the ship. His visibility was also limited to some extent.'

'With a windmilling prop and all of the drag which it created, we were unable to keep up with the formation and became a single straggler on the way home, with heavy fuel consumption on a very deep penetration. It is doubtful whether we could have made Britain. I did a fuel consumption problem after a while and had us running out of fuel about the time we would reach the Channel. It would have been interesting to see whether we could have made it or not. We expected fighter attacks and expected to get knocked down soon because, Hollywood notwithstanding, a single B-17 is mismatched against a flurry of fighters. We made it alone without any fighter cover all the way to a point a few miles east of the Rhine

B-17F-65-BO 42-29723 *Wolf Pack* which was assigned to the 91st Bomb Group at Bassingbourn on 20 April 1943. It later operated with the 546th Bomb Squadron, 384th Bomb Group and 2 BAD at Little Staughton before returning to the USA in June 1944. It was scrapped in December 1945. *(via Mike Bailey)*

before six 109s hit us from the rear. We got two of them for sure and had a probable; a third that was streaking for the ground with heavy smoke pouring out. Since we didn't return to base we couldn't report our two kills and one probable. The 109s plastered us good, knocking out the controls and who knows what else. We bailed out and were captured immediately. Farmers with pitchforks and soldiers were literally waiting beneath for me to land. We spent almost a year in Stalag Luft III.'

John O. Ramsey, 493rd Bomb Group, recalls the 7 October 1944 mission to Merseburg.

'It was our crew's second mission. There was almost a surly silence after the pandemonium that had taken place over the interphone system between the IP, the target area and the

RP (Rally Point). The silence was a state of shock. Never before had we seen a "mighty Fortress" fall from the sky like a game bird caught in the blast of a hunter's shotgun. Others were spiraling down, some trailing white smoke, others, black smoke. Our aircraft, called by some "The Jinx" because of the number of wounded it had brought home and the amount of flak it seemed to attract, took its share this day. One shell exploded just above the fuselage between the nose and the No. 3 engine. Fortunately most of the shell fragments had gone up. The right tyre was ripped open. There were three large holes beneath the co-pilot's seat. Three pieces of flak that entered the right side of the nose compartment tore through the navigator's neatly folded flight

A P-47 Thunderbolt tucks in tight on the wing of a 91st Bomb Group B-17. This P-47 is either seeking the mutual protection of the bomber formation or is escorting them. Thunderbolts were also used as monitor ships to report on practice formation flying over England. *(via Robert M. Foose)*

B-17G *Helena* of the 384th Bomb Group showing the increased frontal gun arrangement used on the 'G' which was designed to counter head-on attacks by the *Luftwaffe*. *(via Robert M. Foose)*

jacket, which was on top of his parachute bag beside his seat and went straight out the left side. The radio operator's steel helmet had a one-inch deep groove in it. He insisted that someone examine him for a bloody wound, which fortunately he did not have, just a very big lump on his head.'

'Our crew had never before seen flak explode so close that we could see the red burst of fire in the centre of the exploding shells. Nor had we smelled cordite mixed with the air and oxygen that we breathed. Black gun smoke slid past our windows. We could hear the crack of the exploding enemy shells over the roar of the forty-eight hundred horsepower of the four engines. All this from an unseen enemy five miles below. No one to turn our twelve machine-guns on. No way to fight back...'

Fred Huston, a bombardier in the 337th BS, 96th Bomb Group, recalls that raids on oil targets at Brux and Merseburg were probably among the worst missions of the war.

'Brux the second time, November 1944, was the first time I saw colored flak. I remember it as red, although there are those who remember it as pale blue. I still think it was red. We were told by someone, probably someone who didn't know either, that it was to call in the fighters. Since we already had more fighters than Custer had Sioux, I thought that it was gilding the lily somewhat to call in something that we already had in abundance. But to me, one fighter was plenty for the day.'

Cold chills over Merseburg

'I am sure that I went to Merseburg something like four or five times. I can also recall a sinking feeling in the pit of my stomach every time the string headed that way. When the string reached Merseburg I had my usual cold chills. But off we went, this time with an IP south of Leuna. We were told that since Hamburg was no longer a target most of the guns had been

moved to the Leipzig area and after the raid I believed them. The First Division led and they missed their IP and came across the target with a quartering head wind and only doing something like a hundred ground speed across the target. They were getting the hell kicked out of them by both flak and fighters. The sky was filled with that peculiar grey green smoke that an exploding airplane left. This was not to say that they didn't have an abundance of guns left for us. We hit highly concentrated flak before we got to the IP, which wasn't the rule and was terribly unsporting of the flak crews. There was a lot of it and it was bouncing us around considerably and we lost a few airplanes before we got on the run properly.'

'We made the bomb run with fewer than the eighteen aircraft we started with and holes kept appearing in various parts of the airplane, like the wings and the bomb bay doors and the nose and everyplace else. Final count was something like four hundred of them in the airplane. Just before bombs away one exploded just above the astrodome, removing it from the airplane and blowing a large and draughty hole in the

B-17Gs of the 91st Bomb Group at Bassingbourn. Left is B-17G-10-VE 42-39996 *Boston Bombshell*, of the 322nd Bomb Squadron, which failed to return with 2/Lt John V. Dunlap and crew on 16 August 1944. The 91st claimed more enemy fighters and lost more bombers than any other 8th AF group. *(via Tom Cushing)*

nose. I knew I was hit because I had a bunch of blood running down my face and out of my glove. My navigator was deaf since a fragment had torn off the earflap of his flak helmet and split his right headset. We left the formation short of our number four engine, and joined some other cripples after the run to get home. Mutual protection was the watchword... the more guns, the fewer Germans. I was resting on the base of the top turret when I stood up and my boot squashed and my leg gave out. That was the last of the battle damage. So yes, I remember Merseburg.'

Posthumous Medal of Honor

It was for his actions on the mission to Merseburg on 2 November 1944 that Lt Robert Femoyer, a navigator in the 447th Bomb Group was posthumously awarded the Medal of Honor. Femoyer's B-17 was rocked by three flak bursts, which showered the aircraft with shrapnel; Femoyer himself was hit in the back and the side of his body, but he refused all aid despite his terrible wounds so that he might navigate the Fortress back to Rattlesden. He was propped up in his seat so he could read his charts, and the crew did what they could for him. It was not until they reached the North Sea that Femoyer agreed to an injection of morphine. He died shortly after the aircraft landed at Rattlesden.

S/Sgt Russell H. Thrift, a B-17 tail gunner in the 493rd Bomb Group, recalls the 31 December 1944 mission to Misburg.

'As we approached the target, we found it to be a perfect day over Misburg, Germany. Just as S-2 had briefed us, there was flak from 6 o'clock to 12 o'clock, and just as thick as London fog. Rockets and flak – everything in the book. Jerry sure was trying to keep us away. The lead squadron went on in and we were right behind heading down that corridor of flak. Then "bombs away." All of the Forts, except those in our squadron had dropped their loads. I guess our squadron's lead bombardier didn't get the target where he wanted it in his crosshairs. When he said, "we're going to take another run," I thought right there that I couldn't have sold my life for a nickel! We banked around

Armament ground crews take out the machine guns from the turrets of B-17G-95-BO 43-38834 *Tutorwolf* of the 332nd Bomb Squadron, 94th Bomb Group, at Bury St. Edmunds (Rougham) following a mission late in 1944. *Tutorwolf* was scrapped in December 1945 at Kingman, Arizona. *(via Abe Dolim)*

Lieutenant Abel L. Dolim, navigator, 332nd Bomb Squadron, 94th Bomb Group. *(Abe Dolim)*

and headed back. Just before we got over the target there was a blinding flash, and all hell broke loose right there in my lap! I was blind and groggy for about a minute, so I can't say exactly what happened. When the dust and splinters cleared, I could see out of one eye. One Fort was going down at 7 o'clock. As for ours, well, flak or a rocket had come up through the armor plating behind me, gone through the ammunition box, hit my right arm, smashed the Plexiglas in front of me, hit my tail sight and knocked it out through the tail window. Boy, that stuff really messes things up when it comes in like that!'

'My guns wouldn't move, and as far as the tail was concerned it was out of action. I remember calling the pilot over the interphone and telling him that I was hit. He sent the radio-operator back with the first aid kit, and

he fixed me up. In the meantime flak had knocked out the oxygen on the right side of the plane and everybody was diving for emergency walk around bottles. When we finally got straightened out, I found I was OK, except for a gash in my left arm and Plexiglas splinters in my eyes. Later, as we were on our way back, and things were running smoothly again, I thanked God for keeping that stuff where it was, another inch or so and, oh well, here I am. I owe a lot to our radio operator (Technical Sergeant Bronislov Sinkus), and the rest of my crew. They took care of things just as though they were participating in a training drill on the ground. The thing I do know; we sure gave "Jerry" a headache that day! They told me that you could see smoke for more than 100 miles from the target.'

Fighting the cold

As the war in Europe drew to a close the B-17s and their crews had to endure the elements as well as enemy opposition. 1/Lt Elwood Samson Jr, co-pilot in Capt David L. 'Doc' Conger's lead crew in the 861st BS, 493rd Bomb Group, recalls the mission to Cologne on 10 January 1945.

'Terrible weather with heavy snow. Our aircraft were de-iced with a spray prior to taxi. However, we had at least one and possibly two holds in take-off position. I was the command pilot. We were the lead group in the bomber stream, about 1,000 aircraft. It was a high priority attack against the Cologne marshalling yards. When we got the green light from the tower our wings were covered with snow and the runway had a good coverage as well. I did not expect it to fly, but being a B-17 it made it! We could not see the star on the top of the wing until the ice had melted at about 20,000 feet.'

In April 1945 Jule F. Berndt, navigator in Lt Rolland B. Peacock Jr's crew in the 850th BS, who had completed a tour of missions in the 490th Bomb Group, reflected on it.

'It almost seems like a bad dream to me now. I just can't visualize myself as having gone through a complete tour of missions in Europe, of having been in England, over Germany, of having flown the ocean and of having crossed it by liner. The memories of those anxious moments over targets, of seeing the face of our

B-17Gs at high altitude leaving their telltale contrails en route to their target. *(via Mike Bailey)*

wounded ball turret gunner, of the time over Duisburg and Merseburg when there just didn't seem to be a plausible excuse for emerging unscathed from the clouds of flak that we entered – all these now seem like parts of one of those bad nightmares that are hard to reconstruct after you wake up. The long hours of thinking before falling asleep on the night before a mission and the anxious moments spent contemplating the thought of dying so young – worrying about such things not just for yourself but also for your parents, who you knew were praying for your safety back home. All this has become just a part of the past, and I am here now, today, whole, alive and writing about it. It is almost too good to be true.'

Not everyone who flew in B-17s remembers the Fortress with total affection. Lt Abel L. 'Abe' Dolim, a navigator in the 332nd Bomb Squadron, 94th Bomb Group, recalls:

'Men seem to either love or hate the machines they fly. My first co-pilot had been pulled out of B-25 phase training and transferred to B-17s. He never got over his love affair with the B-25 and accepted his lot with great bitterness. He was like a passionate young lover torn from the bosom of his true love and banished to exile among strangers by harsh and disapproving parents.'

'My affair with the B-17 was like an infatuation with a comely young wench. How was I to know that she would eventually lead

B-17Gs of the 351st Bomb Group en route to Berlin on 26 February 1945. *(USAF)*

me down the primrose path to perdition? She was lover, mother, and comforter...all that a young swain hopes for in his loved one. Then alas, as in all earthly relationships, reality intruded and destroyed my dreams. I began to see her faults, her aberrations, and all the petty deceits that marred her beauty. I tried to come to terms but the advantages were all on her side because of her great power over me... it was a true power of life and death... and it became in time a love-killing death blow from which I never recovered. Love turned sour, I began to watch her every move. My nerves took a beating. She developed suicidal tendencies... I had the feeling she wanted me to share in her death wish. I was terrified and fiercely resolved

to escape her clutches the very instant she made any attempt to involve me in her wild schemes. The strain eventually got to me and I longed for an end to our deadly relationship. The last weeks and days were an agony. Finally it was all over, I was FREE, home safe, and I didn't give a damn if I ever laid eyes on the old whore again.'

'The B-17 – I am not ashamed to say that in 1944 and 1945 I regarded her as a slow, flammable old lady who was really out of sorts among fast young friends, and who became my potential enemy every time I flew a combat mission. Had it not been for the long-range escort fighter, the *Luftwaffe* would have driven us from the skies over Germany.'

B-17Gs of the 490th Bomb Group en route to their target amid bomber contrails, with the higher crossing contrails marking the comforting presence of 8th AF fighter escorts. *(via Mike Bailey)*

B-17G-70-DL 44-6975 of the 381st Bomb Group which bellied in at Ridgewell on 15 February 1945. *(Joseph Minton via Larry Goldstein)*

In May 1945 at the end of hostilities in Europe, hundreds of aircraft wrecks littered the continent. B-17G-30-VE 42-97781 BN-O *Eight Ball III*, 359th BS, 303rd Bomb Group, came down on the edge of the Apeldoorn-Diemen Canal at Eerbeek, Holland, on 2 November 1944. Nine of Lt Jack T Davis' crew were made PoW and one man was KIA. *(Brouwer, via Ab A. Jansen).*

USAAF B-17 Medal of Honor Awards
1942–44

DATE	RECIPIENT	USAAF UNIT	MISSION
7 August 1942	Captain Harl Pease Jr+	19th BG/5th AF	Rabaul, New Britain
18 March 1943	1/Lt Jack Mathis*	303rd BG/8th AF	Vegesack, Germany
1 May1943	S/Sgt Maynard H. Smith	423rd BS/306th BG/8th AF	St Nazaire, France
16 June 1943	2/Lt Joseph R. Sarnoski*	43rd BG/5th AF	Buka, Solomon Islands
16 June 1943	Major Jay Zeamer Jr	43rd BG/5th AF	Buka, Solomon Islands
26 July 1943	Flt Off John C. Morgan	92nd BG/8th AF	Kiel, Germany
20 December 1943	T/Sgt Forrest L. Vosler	303rd BG/8th AF	Bremen, Germany
20 February 1944	1/Lt William R. Lawley	305th BG/8th AF	Leipzig, Germany
20 February 1944	Sgt Archibald Mathies*	351st BG/8th AF	Leipzig, Germany
20 February 1944	2/Lt Walter E. Truemper*	351st BG/8th AF	Leipzig, Germany
11 April 1944	1/Lt Edward S. Michael	305th BG/8th AF	Brunswick, Germany
23 June 1944	2/Lt David R. Kingsley*	97th BG/15th AF	Ploesti, Romania
2 November 1944	2/Lt Robert E. Femoyer*	711th BS/447th BG/8th AF	Merseburg, Germany
9 November 1944	1/Lt Donald J. Gott*	729th BS/452nd BG/8th AF	Saarbrucken, Germany
9 November 1944	2/Lt William E. Metzger*	729th BS/452nd BG/8th AF	Saarbrucken, Germany
24 December 1944	Brig-Gen Fred W. Castle*	4th BW/8th AF	During Battle of the Bulge

+ The Japanese executed Captain Pease in October 1942
* Posthumous Award

5. Fortress at War: Versions and Variants

Results of European combat experience and lessons learned in the Pacific were incorporated into the extensively improved B-17E, which was ordered on 30 August 1941. The rear fuselage from the radio compartment on was extensively re-designed to provide more space for the gunners, and greatly enlarged tail surfaces gave better control and stability for high-altitude bombing. A new tail gun position was added, with two .50-calibre Browning M2 machine-guns fired by the gunner from an uncomfortable half-kneeling, half-sitting position.

The ventral bathtub was deleted on the first 112 B-l7Es and replaced with a new Bendix power-operated gun turret with twin .50s fired by a gunner using controls and a periscope sighting arrangement in the fuselage. The turret proved troublesome to operate and was subsequently replaced with the Sperry ball turret with the gunner squeezed inside. A Sperry electrically operated turret with twin .50s was installed behind the cockpit just in front of the bomb-bay. The single .30-calibre machine-gun in the nose was retained, as it was thought no enemy fighter pilots would attempt a head-on attack with such high closing speeds between fighter and bomber.

Large orders

Boeing received orders for 812 B-17Es but material shortages delayed production. The 'E' did not make its maiden flight until 5 September 1941, four months behind schedule. About 100 B-l7Es had been delivered to the Air Corps by the time of the Japanese attack on Pearl Harbor on 7 December 1941. Fifty more on hand were the older 'C' and 'D' models. A handful of bomb groups fought the Japanese in the Philippines and Java and the survivors retreated to India.

Early in 1942 the AAF became interested in improving the powerplant of the B-17 by using liquid-cooled Allison V-1710-89 engines of 1,425 hp. Boeing was fully committed to B-17 production and development work on the B-29 Superfortress, so on 10 July 1942, Lockheed Vega was contracted for three conversions of B-l7Es using the new engines. 41-2401, the ninth production Boeing B-17E model, was loaned to Lockheed as the pattern aircraft for the BVD programme. Though known as the XB-38, Lockheed assigned its own model number, V-134-1. Lockheed Vega installed the four Allison V-1710-89 powerplants, removed the oil coolers from the wing leading edges and put them in the nacelles, and installed new semi-circular intakes. Coolant radiators were positioned in the leading edges of the wing between the nacelles. The XB-38 first flew on 19 May 1943 and early indications were that the performance was better than that produced by the Cyclones, with a top speed of 327 mph at 25,000 ft at a gross weight of 58,000 lb. However, the service ceiling was only 29,100 ft compared to 36,000 ft which was attained by the B-17E. The XB-38 had accumulated just twelve hours' flight time when, on 16 June 1943, it got into trouble. Test pilot Bud Martin levelled off at 25,000 ft and opened the throttles

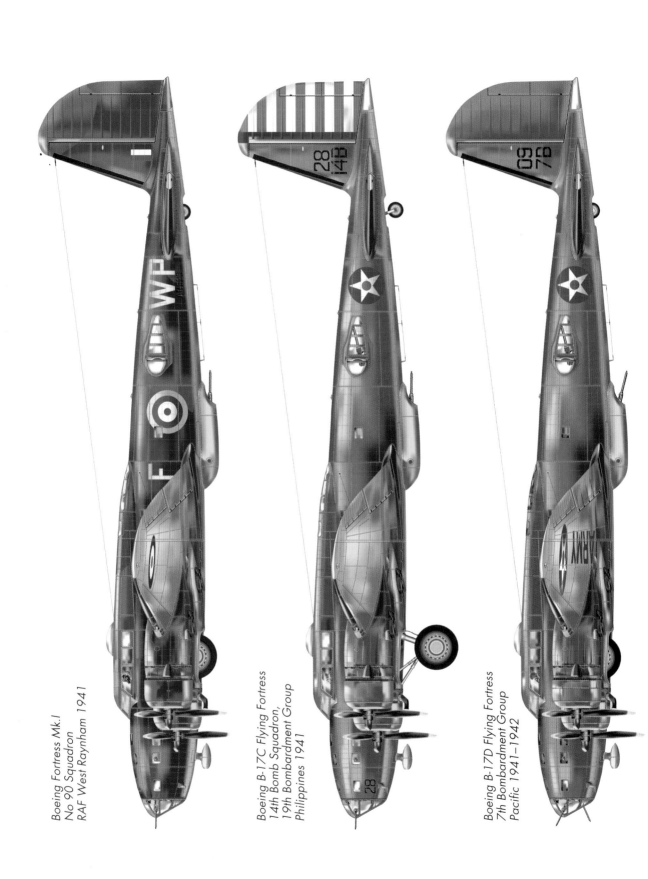

Boeing Fortress Mk.I
No 90 Squadron
RAF West Raynham 1941

Boeing B-17C Flying Fortress
14th Bomb Squadron,
19th Bombardment Group
Philippines 1941

Boeing B-17D Flying Fortress
7th Bombardment Group
Pacific 1941–1942

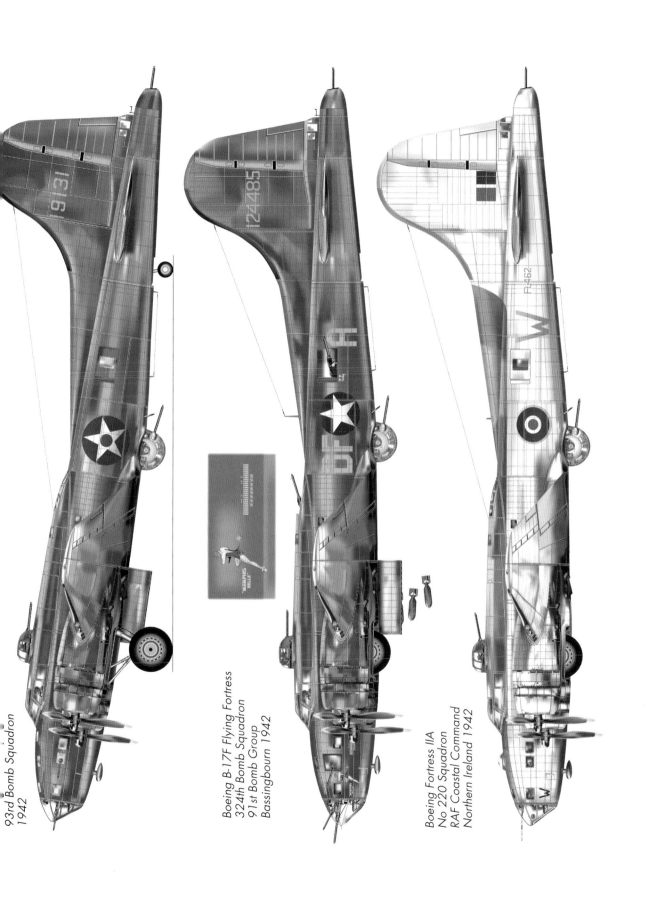

93rd Bomb Squadron
1942

Boeing B-17F Flying Fortress
324th Bomb Squadron
91st Bomb Group
Bassingbourn 1942

Boeing Fortress IIA
No 220 Squadron
RAF Coastal Command
Northern Ireland 1942

B-17Es on the line at Boeing Field in Seattle. The nearest aircraft is B-17E 41-2393, which was delivered to Wright Field on 3 October 1941 and went on to serve in Newfoundland. Starting with the 113th B-17E, the remotely controlled Bendix under-turret seen here, which the gunner operated remotely using a periscope sight, was replaced by the manned Sperry ball turret. *(Boeing)*

for a full power run. Almost immediately, No. 3 engine burst into flames. The extinguisher failed to put out the fire and Martin and all but one of the crew parachuted to safety before the XB-38 crashed near Tipton, north of Bakersfield, California. No further Allison-engined Fortresses were converted and in any event, the Allison powerplant was much in demand for powering several types of fighter aircraft, including the P-38 Lightning and Curtiss P-40.

Another B-17E conversion was the C-108 transport. The first was B-17E 41-2593, which was fitted out as a flying office for General Douglas MacArthur. A B-17F nose cone was fitted, along with a forward firing .50-calibre machine-gun. The single XC-108A was another B-17E converted to a freighter by stripping the interior and installing a large upward-opening cargo door. While this version did not enter production, the cargo door was later installed on a few CB-17s and fitted on some postwar surplus B-17s, which were used as freighters. The single XC-108B was B-17F 42-30190 converted to an airborne tanker to ferry fuel over the Himalayas, or 'the Hump', as it was known, from India to China in 1944.

Glide bomb tests

B-17E 41-2590 was used to test experimental GB-1 *Grapefruit* glide bombs, which were developed by fitting small wings and an empennage to a 2,000 lb GP M34 bomb, and launched from the external wing racks of B-17s. On 28 May 1944 the 384th Bomb Group, 8th AF, launched 116 GB-1 bombs on targets at Cologne, but none even hit the city. The weapon did not see widespread use. The GB-4 *Batty* combined radio-control and visual-observation version was produced using a GB-1 with the *Azon* 'stand-off' glider bomb. This

B-17E 41-9141 in RAF-style camouflage scheme, and B-17E 41-9131 in US Army camouflage and markings, from the second production batch, in flight in America. The B-17E in the foreground was built for Great Britain with standard British camouflage, but was taken over by the US Army and flown in its original camouflage with US stars painted over the RAF roundels on both wings. (Boeing)

consisted of a 1,000 or 2,000 lb GP bomb with radio-controlled movable tail fins. The GB-8 version was controlled visually by rudimentary television. On 24 April 1944, five B-17s of the 301st Bomb Group in Italy, equipped with Azon remotely guided bombs, made the first of two such attacks on the Ancona-Rimini railway line. A second raid was made on the same railway on 29 April. Both the GB-4 and GB-8 were launched experimentally from B-17s in the last few months of the European war, but were largely unsuccessful.

The B-17 continued to operate in the Pacific Theatre until 1943. Beginning in March 1942, the first of 45 B-17Es known as Fortress IIAs was delivered to Britain. Many gave sterling service in RAF Coastal Command where they helped close the mid-Atlantic 'gap'. The first B-17Es of the 97th Bomb Group landed in Britain in July 1942 and this unit flew the first American Fortress mission on 17 August. After 512 B-17Es had been built the remaining 300 aircraft on the contract were converted to B-17F production standard.

Mass production

The 'F' was the first Fortress model to enter really large-scale production. The BVD pool was created when Boeing agreed to let Lockheed Vega at Burbank, California and Douglas Aircraft at Long Beach build the B-17F under licence. Lockheed had first approached Boeing about building the B-17 under licence in April 1941, when it feared that gathering war clouds would limit the need for passenger aircraft in which the Vega plant specialised. Lockheed Vega turned out the first B-17F on 4 May 1942. Boeing followed on 30 May, and Douglas produced its first B-17F on 9 June.

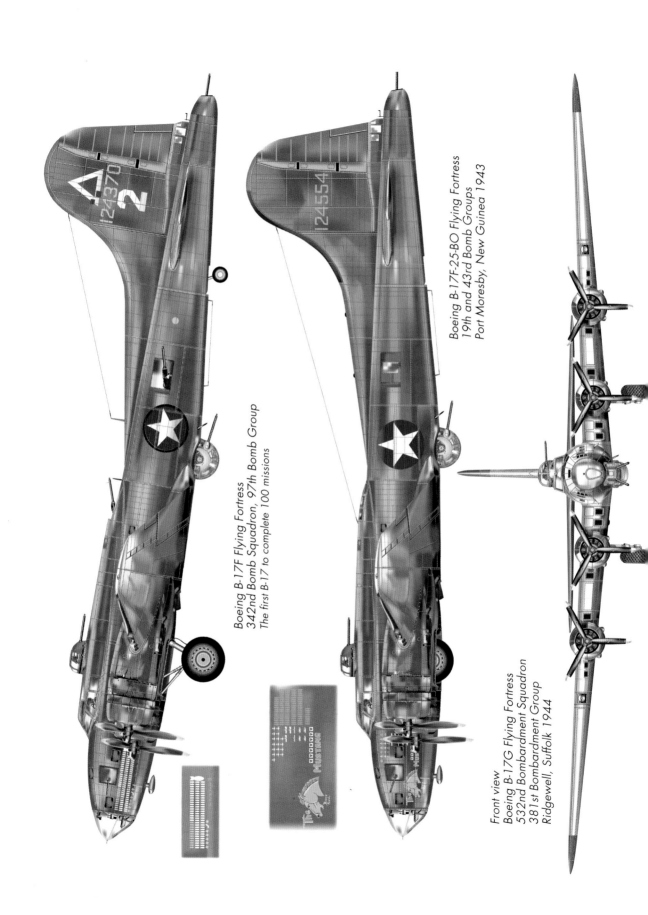

Boeing B-17F Flying Fortress
342nd Bomb Squadron, 97th Bomb Group
The first B-17 to complete 100 missions

Boeing B-17F-25-BO Flying Fortress
19th and 43rd Bomb Groups
Port Moresby, New Guinea 1943

Front view
Boeing B-17G Flying Fortress
532nd Bombardment Squadron
381st Bombardment Group
Ridgewell, Suffolk 1944

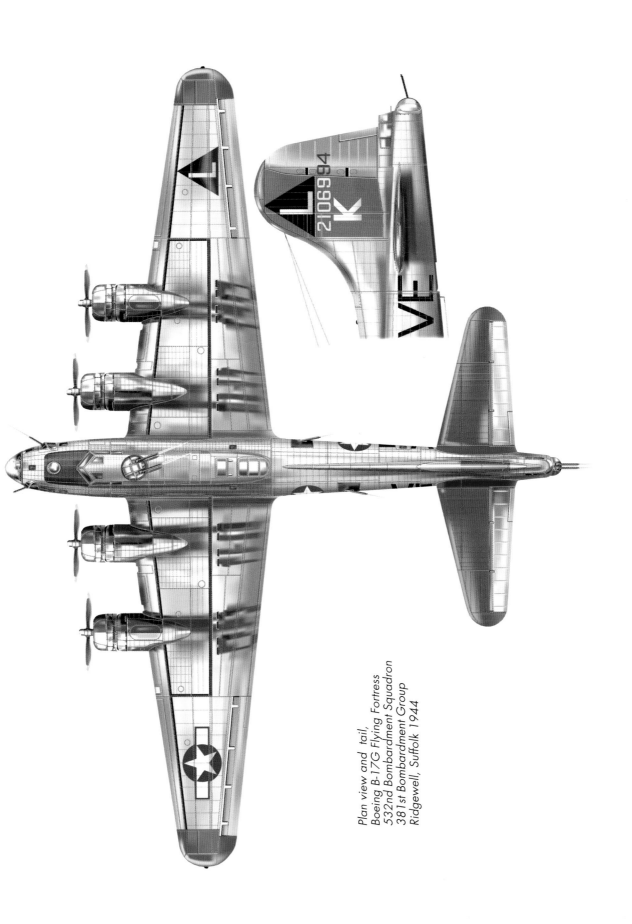

*Plan view and tail,
Boeing B-17G Flying Fortress
532nd Bombardment Squadron
381st Bombardment Group
Ridgewell, Suffolk 1944*

Outwardly, the 'F' appeared to be similar to the B-17E, except for a frameless Plexiglas nose cone, which gave the bombardier better all-round visibility and increased the overall length of the Fortress from 73 ft 9.75 in for the B-17E to 74 ft 8.9 in. Unfortunately, the new nose cone, which was fitted with duplicate sockets for a single .30 calibre gun to fire from the upper quarter or lower right areas of the nose, was too weak to support a .50 calibre machine-gun. Field engineers had to strengthen the area when the B-17Fs reached the war theatres. Problems with the bombardier and navigator getting in each other's way during firing of the guns was improved slightly by enlarging and staggering the side nose gun windows. AAF modification centres bulged the side windows outward to permit the nose gunners a wider forward firing sweep. These bulged nose guns, known as 'cheek' guns, eventually became standard.

Internal redesign

In fact, no fewer than 400 changes and modifications were incorporated in the B-17F design, most of them being carried out on the Boeing, Vega and Douglas production lines. At Boeing, beginning with block No. B-17F-1-BO and starting with serial No. 41-24340, bomb load was increased, the bomb racks were redesigned, and additional armour plate added. Beginning with B-l7F-5-BO/41-24390,

armour plate was installed in the waist gun areas. Starting with B-17F-10-BO/41-24440 the tailwheel was strengthened, while B-17F-20-BO/41-24504 saw sockets for the .30-in guns removed from the nose area. B-17F-27-BO/41-24585 introduced a strengthened undercarriage drag strut, and beginning with B-17F-30-BO/42-5050, external bomb racks were fitted and the ball sockets in the side windows were removed. B-17F-50-BO/42-5350 introduced continuous flexible ammunition feeds for the waist guns, while heavy-duty brakes were fitted from the 610th aircraft onwards. With B-17F-55-BO/42-29467, came cheek guns. Beginning with B-l7F-75-BO/42-29832, the radio compartment received a .50-in machine-gun and .50-calibre nose guns. With B-17F-80-BO/42-29932 came 'Toyko tanks' in the outer wing sections. The 540-gallon 'Tokyo tank'was so called because someone said that with these the Fortress could fly to Japan. Two tanks in each outer wing gave the B-17F a total of 2,780 gallons as standard, or 3,630 gallons with two 410-gallon tanks in the bomb bay for ferry flights. The 'Tokyo tanks' were vented to allow for variations in the air pressure with changes in altitude. With the B-l7F-95-BO, starting at 42-30232, the external wing bomb racks were deleted and the left and right tailplanes made interchangeable. Beginning with B-17F-110-BO/42-3 0532, a flexible ammunition chute was fitted for the tail guns. On B-17F-120-BO,

This B-17E, 41-2401, was modified by Lockheed Vega to include four Allison V-1710-89 liquid-cooled V-12 engines of 1,425 hp each and redesignated XB-38. It flew for the first time on 19 May 1943 and proved faster than Wright-engined B-17s, though it was not as good at altitude. Development was not proceeded with after it crashed on 16 June following an engine fire in flight. *(Lockheed)*

B-17Fs on the Douglas, Long Beach production line. Douglas manufactured 3,000 Flying Fortresses, the bulk of which were 'G' models completed in 1944 and 1945. *(McDonnell Douglas)*

starting with 42-30 732, a life raft inspection window was added.

At Douglas Long Beach, beginning with B-17F-1-DL/42-2964, double windows were fitted in the pilots' windscreen; and recognition lights added to the underside of the fuselage. Starting with B-17F-10-DL/42-2979, provision was made for carrying 1,600-lb armour-piercing bombs; while the fittings for .30-in machine-guns was deleted. Like the Boeing model, armour was added to the waist gunners' area. Beginning with B-17F-15-DL/42-3004, a .50-in gun and a 250-round ammunition box were provided for use in the radio room. The waist gunners received continuous flexible ammunition feeds for their guns; provision was made for two .50-in nose guns, and an astrocompass bubble was also added to the nose just in front of the windscreen. Block No. B-17F-35-DL (beginning with 42-3189) saw the introduction of a windscreen wiper and strengthened mainwheels. B-17F-45-DL/45-3504 introduced a Bendix chin turret installation just under the bombardier's seat in the nose. Gun direction and firing were controlled through a firing yoke mounted in the right side of the forward nose.

At Lockheed Vega, beginning with B-l7F-1-VE/42-5705, a recognition light was added and starting with B-17F-10-VF/42-5725, improved fuel tank vents were installed. Beginning with the B-17F-l5-VE/42-5745, an astrocompass bubble was added to the nose just in front of the windscreen. B-17F-20-VE/42-5765 saw the external bomb racks deleted, and starting with B-17F-30-VE/42-5855, 'Tokyo tanks' were installed, the main wheels were strengthened and heavy-duty brakes installed.

More engine power

Powering the new model were Wright R-1820-97 Cyclones, which were rated at 1,200 hp for take-off and 1,380 hp war emergency power. These more powerful engines, which gave the B-17F a top speed of 325 mph, needed improved propellers and so Hamilton Standard broad chord 'paddle' propeller blades were fitted to absorb the extra output and increase

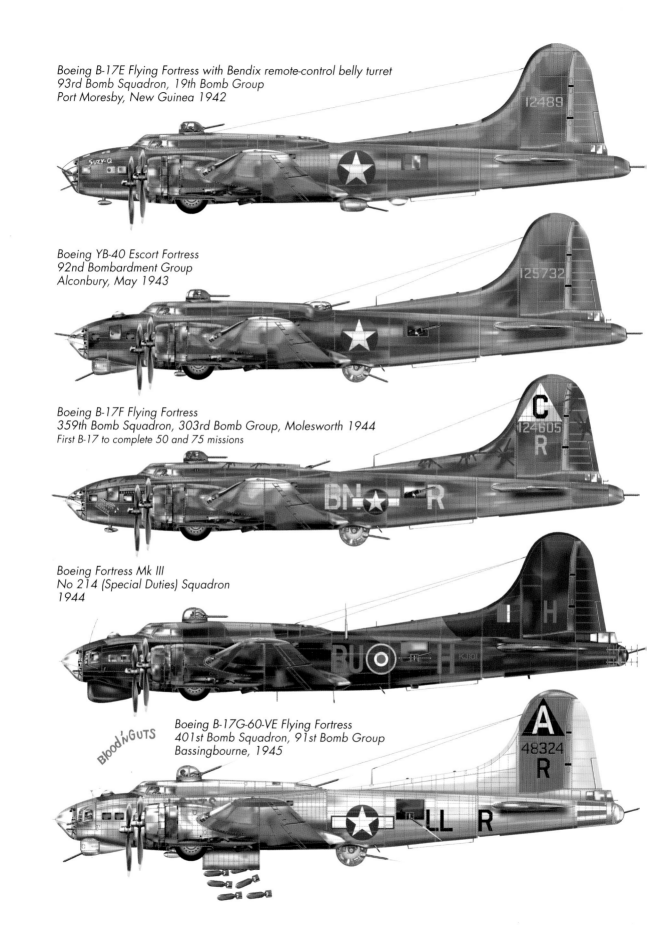

Boeing B-17E Flying Fortress with Bendix remote-control belly turret
93rd Bomb Squadron, 19th Bomb Group
Port Moresby, New Guinea 1942

Boeing YB-40 Escort Fortress
92nd Bombardment Group
Alconbury, May 1943

Boeing B-17F Flying Fortress
359th Bomb Squadron, 303rd Bomb Group, Molesworth 1944
First B-17 to complete 50 and 75 missions

Boeing Fortress Mk III
No 214 (Special Duties) Squadron
1944

Boeing B-17G-60-VE Flying Fortress
401st Bomb Squadron, 91st Bomb Group
Bassingbourne, 1945

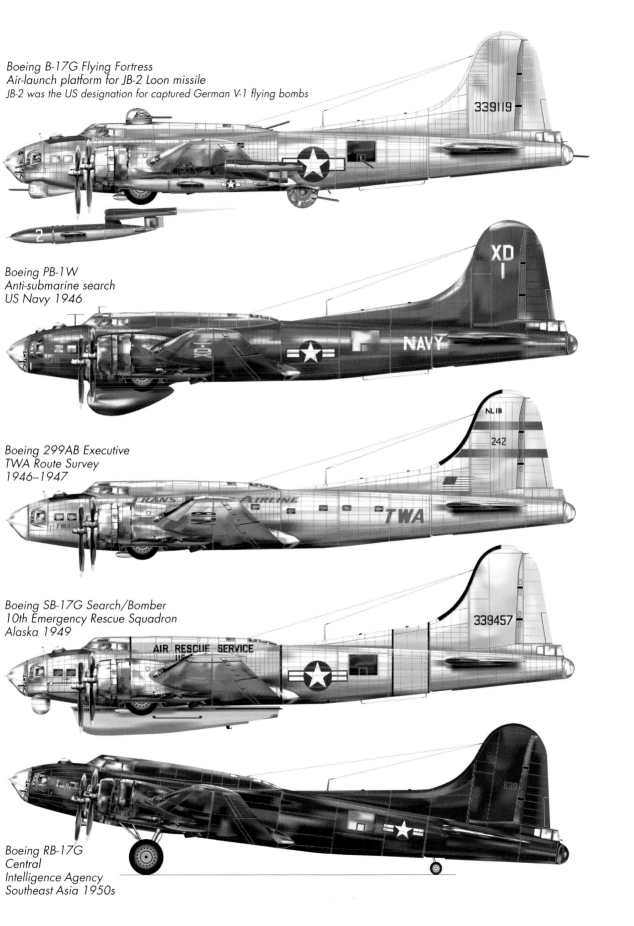

Boeing B-17G Flying Fortress
Air-launch platform for JB-2 Loon missile
JB-2 was the US designation for captured German V-1 flying bombs

Boeing PB-1W
Anti-submarine search
US Navy 1946

Boeing 299AB Executive
TWA Route Survey
1946–1947

Boeing SB-17G Search/Bomber
10th Emergency Rescue Squadron
Alaska 1949

Boeing RB-17G
Central
Intelligence Agency
Southeast Asia 1950s

Fortress IIA FK184 (B-17E 41-2513) of No. 220 Squadron, RAF Coastal Command on patrol in 1942. The B-17Cs and Es supplied to Great Britain were known as IIAs because they were existing aircraft diverted for maritime patrol use and the 'F' had not yet been built. *(via Mike Bailey)*

'bite' at altitude. The leading edge of the engine cowlings had to be re-shaped and shortened so that the blades could clear the cowling when feathered. Range was now 4,220 miles. The maximum bomb-load was 9,600 lb, but the normal combat load was nearer 4,000 lb. The first B-17F-1-BO was delivered to the Army Air Force (the Air Corps had become the US Army Air Force in June 1941) on 30 May 1942 and after August that year, 19 B-l7Fs went to RAF Coastal Command as the Fortress II. Originally, the British had signed a lend-lease contract in June 1941 for 300 Fortress IIs for the RAF but these aircraft were diverted to the USAAF when Britain decided not to use the B-17s as bombers but only for maritime operations.

At the beginning of 1943 the 8th Air Force in Britain comprised just four B-17F groups totalling some 200 aircraft. On 27 January VIIIth Bomber Command bombed Germany for the first time when B-17Fs of the 306th Bomb Group attacked Wilhelmshaven. During mid-July hundreds of B-17Fs were used in a week of attacks beginning on 24 July which came to be known as 'Blitz Week'. Despite high losses the B-17s were despatched to Schweinfurt and Regensburg a month later, on 17 August. Sixty B-17s were lost and many others written off in crashes in England; an unacceptable 19 per cent loss-rate. On 6 September 45 B-17s failed to return from a raid on Stuttgart. The worst day, 'Black Thursday', occurred on 14 October, when another 60 B-17s were lost, again on a mission to Schweinfurt.

Boeing-Vega-Douglas

Altogether, B-17F production totalled 3,400. Boeing built 2,300 and starting in July 1942, 600 and 500 each were delivered by Douglas and Lockheed Vega respectively, in new factories built specially for the purpose. Many modifications were phased in during assembly and considerable confusion developed in the war zones when it was discovered that although each manufacturer used the same block numbers in production, not all the aircraft were similarly equipped. Differences often occurred in nose armament but generally the BVD pool was a great success and greatly speeded the flow of Fortresses to the war zones.

The F-9s were B-l7Fs and B-17Gs fitted with various high-altitude cameras for long-range reconnaissance work but initially retained their defensive armament. Differing camera arrangements resulted in designations to F-9C to a total of approximately 100. In 1948, surviving F-9s still used for photography were redesignated FB-17G or RB-17G.

Meanwhile, in August 1942 it was suggested that bombers be accompanied by 'destroyer-

B-17F-1-BO 41-24341 was modified as the XB-40 escort aircraft by Vega. It is seen here at Burbank before its first flight on 10 November 1942. Douglas modified twenty Vega-built B-17Fs as YB-40 service-test models. Seven YB-40s made their operational debut on 29 May 1943 with the 92nd Bomb Group and flew missions until the end of July 1943. Vega received no order for its proposed V-140 production version. *(Boeing)*

escort' aircraft, since existing fighter escorts did not possess the range to accompany the bombers deep into Germany. One proposal was a heavily armed Flying Fortress. In November 1942 Lockheed took 41-24341, the second Boeing-built B-17F, and converted it into the V-139-3 or XB-40 'destroyer escort' as it was better known. Two additional gun turrets; a Martin in place of the radio compartment guns and a Bendix chin turret were fitted, and the single waist guns were replaced by twin .50s, making a total of 14 guns in total. Both the single waist guns were replaced by pairs of powered .50s, and the tail guns were hydraulically boosted for improved control. Early B-17Fs were loaded with 3,400 rounds of .50 and 500 rounds of .30 calibre ammunition but the XB-40 was designed to carry 11,275 rounds of ammunition, rising to 17,399 rounds for short-range missions.

In all, 22 pre-production YB-40s were built. They were designed by Boeing, built by Lockheed Vega on the B-17F contract, and converted by Douglas Aircraft at Tulsa, Oklahoma, although they were identified as Vega-built aircraft. In January 1943 twelve were flown to England to be used by the 327th Bomb Squadron, 92nd Bomb Group at Alconbury. The first raid involving YB-40s took place on 29 May 1943 when four accompanied the

Fortresses to St Nazaire. It was evident after the first few weeks of operations that the YB-40 did not add materially to the combined firepower of a group and it could not protect stragglers from a concentrated enemy fighter attack. With a normal cruising speed less than that of the B-17Fs, the YB-40s had to put on the power just to keep up and in doing so, they used more fuel than the bombers. On the return from a target the B-17s weighed up to 6,000 lb less having dropped their bombloads, but the YB-40 still weighed almost the same as on the outward leg. The YB-40 was only used in very small numbers and flew its final mission on 28 July. The final four YB-40s were never delivered as such; instead they were modified and used as TB-40 training aircraft in the US.

Nose armament

One tangible benefit that resulted from the abortive YB-40 project was the chin turret, which was adopted in the last 86 Douglas-built B-l7Fs and the final production version, the B-17G. The early B-17G retained the nose window configuration of the early B-17F. All three manufacturers added bulged cheek windows at various stages of production. Beginning with the Boeing B-17G-50-BO, the waist gun positions were staggered to allow the gunners more freedom of movement in combat. The United Air Lines Modification Centre at Cheyenne, Wyoming designed an all-new tail turret with enlarged windows and a reflector gunsight in place of the ring and bead sight. The *Cheyenne* turret provided a completely redesigned gunner's enclosure, with much more flexible gun elevation. It was installed during various stages of production at all three B-17G plants.

As with the B-17F, there were many myriad improvements throughout B-17G production. At Boeing, beginning with B-17G-1-BO/42-31032, a Bendix chin turret was added, the bombardier's windscreen wiper was removed; and a revised transparent nose cone was fitted. Starting at B-17G-5-BO/42-31132, provision was made for additional oxygen bottles to be carried on the ball turret suspension frame. With B-17G-l5-BO/42-31332 came a modified A-2 ball turret and 600-round ammunition boxes. B-17G-20-BO/42-31432 introduced the Type A-2 ball turret with oxygen swivel. From B-17G-35-BO/42-31932, Type B-22 turbo superchargers were installed. A bulbous elevator trim tab was also fitted. Starting with B-17G-40-BO/42-97058, external access was permitted to the life raft compartment, and on B-17G-50-BO/42-102379 enclosed waist gun positions with K-6 mounts made their appearance. Starting from B-l7G-60-BO/42-102744, the British *Lorenz* blind approach equipment was deleted; and cheek guns were installed. The B-17G-70-BO/43-37674 saw the re-installation of the trailing wire static discharger at the tailwheel; and fore and aft radio compass antennas were fitted. On B-17G-75-BO/43-37874, the B-10 type replaced the Type B-17 bomb shackles.

Cabin heating

Other improvements included modified microphone switches on the hand-held guns and the bombsight, as well as the addition of parachute static lines. Beginning with B-l7G-80-BO/43-38074 the Fortress had a hot air cabin heating system; chin turret with hand gun chargers; and external life raft release handles. B-17G-85-BO/43-38374 saw the installation of an enclosed radio operator's gun. A popular improvement was the 'formation stick', an electric power boost for the control column. Activated by a pistol grip, it relieved the workload needed to hold a heavy B-I7 in formation and was introduced early in 1945.

Lockheed modifications

At Lockheed Vega starting with B-17G-l-VE/42-39758, the bombardier's windscreen wiper was removed and the life raft compartment received inspection windows. The instrument panel was cut-out for an AN-5735 turn indicator. Beginning with B-17G-100-VE/42-39958, the British *Lorenz* blind approach radio was installed as was the A-2A ball turret. B-17G-15-VE/42-97436 saw the introduction of Type A-9 wingtip landing lights. Beginning with B-17G-25-VE/42-97636, Type B-22 turbo superchargers were fitted. On B-17G-30-VE/42-97736, ammunition boxes for cheek guns were fitted and starting with B-17G-35-VF/42-97836, cheek guns were installed. On B-17G-45-VE/44-8001, Type B-10 bomb shackles were used, while the bomb-bay fuel tank attachments were deleted; and parachute static lines introduced at the navigator's hatch and the main entry door. Beginning with B-17G-50-VE/44-8101, a trailing wire static discharger was installed at the tailwheel, Type K-6 gun mounts were used in the waist areas and hot air cabin heating was installed. B-17G-55-VE/44-8201 had chin turret hand chargers, external life raft release handles, an enclosed radio compartment gun and the tail guns were modified to incorporate the N-8 sight. Beginning with B-17G-60-VE/44-8301, a modified microphone switch was used on the hand-held guns. Starting with B-17G-65-VE/44-8401, wiring provision was made for the Type N-8A sight on the cheek guns and an enclosed radio operator's gun was installed. The 'formation stick' was introduced into B-17G-80-VE blocks early in 1945.

At Douglas Long Beach, starting with B-17G-15-DL/42-37804, 600-round ammunition boxes were used. On B-l7G-20-DL/42-37894, provision was made for a removable panel in the windscreen. Starting from B-17G-25-DL/42-37989, the waist guns were enclosed, and cheek guns installed. B-17G-30-DL/42-38084 saw the Type A-2A ball turret fitted while B-17G-35-DL/42-106984 saw the partial replacement of B-7 bomb shackles with the Type B-10. Also, the

B-17G-30-DL 42-38213 (olive drab) and B-17G-35-DL 42-106984 (natural metal finish) at the Douglas Long Beach factory. 42-106984 did not leave the US, while 42-38213 was assigned to the 20th Bomb Squadron, 2nd Bomb Group, 15th Air Force. It was lost on 7 July 1944 on the mission to Blechhammer. Douglas-built 'G' models were the first to become operational, the USAAF receiving its first 'G' on 4 September 1943. Leaving the bomber unpainted speeded up production, and the smooth metallic finish reduced drag, increasing overall flight performance. Altogether, Douglas built 2,395 B-17Gs. (McDonnell Douglas)

bomb hoist pulley and cable assemblies were deleted. Starting with B-17G-40-DL/44-6001, high-speed turbo-chargers and K-6 gun mounts were introduced. From B-17G-45-DL/44-6126 onwards, B-10 bomb shackles were used throughout and the bomb-bay fuel tank attachments were deleted. A trailing wire static discharger was fitted at the tailwheel and external life raft handles were modified. Beginning with B-17G-50-DL/44-6251, the chin turret hand chargers were modified. The 'formation stick' was introduced in block B-17G-70-DL aircraft.

'G' model flies

The B-17G flew for the first time on 21 May 1943 and first began to equip bomb groups of the 8th Air Force in England in September 1943. The arrival of the P-51 Mustang enabled 1,000-plane raids to be flown deep into Germany.

During March 1944 the first American raids on Berlin took place and by April the 8th AF could call upon almost 1,000 B-17Gs; the number was to double by August that year.

Although in the main the B-17s dropped conventional HE and occasionally launched glide bombs, some Forts actually became 'glide bombs' in their own right. In the late summer of 1944, a number of war-weary 8th AF B-17s were used at Fersfield, Suffolk, as BQ-7 'drones' in a project code-named 'Project A' or *Aphrodite*. The US Navy part of the project was called *Anvil*. These drones were stripped of all armour plate, guns and equipment, fitted with radio controls, and packed to the gunnels with up to 20,000 lb of *Torpex*, a naval high explosive, or 10 tons of *RDX* explosive.

A two-man crew would manually take the drone off and put it on course for a heavily reinforced enemy target, such as a V-2 site,

B-17s were used to deliver a number of special weapons during the war. B-17G-95-BO 43-38832 *Irene* of the 306th Bomb Group at Thurleigh is seen here with a rocket-boosted, concrete-piercing *Disney* bomb under its wing. The bomb was designed to destroy hard targets like U-boat pens. *(USAF)*

before bailing out over England. At this point the drone was guided to its target by a control aircraft, usually a B-34 (an ex-RAF Lockheed Ventura). The project was abandoned after several crashes, near misses (and the loss of a US Navy *Anvil* drone which killed Lt Joe Kennedy Jr and his co-pilot on 11 August 1944, when their PB4Y Liberator detonated prematurely). On 4 August, for instance, four *Aphrodite* B-17s were despatched against a V-1 site in the Pas de Calais, each loaded with 10 tons of *RDX*. None reached their intended targets: two were shot down, and the other two crashed and exploded, one after the crew had baled out. Two days later, two more drones were launched. One crashed into the Channel, and the other exploded just after the crew had baled out over East Anglia.

Disney bombs

Another device for destroying heavily protected German targets, the forerunner of the 'bunker buster' bomb, came courtesy of the British. Royal Navy Captain Edward Terrell developed a 4,500lb bomb, powered by a rocket motor in the tail, which was designed to

penetrate 20 ft of concrete before exploding. Too big to be carried in the bomb bay of a B-17, two of these so-called *Disney* bombs could be carried under each wing of a B-17. *Disney* bombs were first used on 10 February 1945 when nine B-17s of the 92nd Bomb Group dropped eighteen of them on E-boat pens at Ijmuiden in Holland. One hit was recorded, and further trials were ordered, but the Allies' sweeping victories in the Low Countries and the vast distance to suitable targets in Norway brought the *Disney* missions almost to an end.

Among the most modified Fortresses flying in World War II were the Fortress IIs and IIIs used by Nos. 214 and 223 Squadrons, RAF. They were part of 100 (Bomber Support) Group, which served in the very valuable radar-countermeasures (RCM) and electronic intelligence (ELINT) roles. The primary duty of the 100 Group 'jammers' was to reduce the escalating losses being suffered by RAF Bomber Command over the Reich at night, by interfering with enemy night fighter radio and radar transmissions. In RAF 100 Group service, engine exhaust flame dampers were fitted, the chin turret was replaced by a fairing for the

An electronic warfare Fortress III, KH999 BU-W of No. 214 Squadron RAF 100 Group. Note the prominent Jostle transmission mast behind the radio room, nose radome containing the *H2S* scanner, and the rear mounted *Airborne Grocer* and *Dina* aerials. *(via Eileen Boorman)*

H2S radar scanner, and the ball turret, bomb aiming and release gear were removed, although the bomb racks were retained. *Gee* navigation equipment was also fitted.

Electronic warfare

Early in 1944 the first 100 Group B-17s were fitted with *Airborne Cigar* or *ABC* pending installation of *Jostle IV* VHF R/T jamming equipment. *ABC* was a device consisting of six scanning receivers and three transmitters designed to cover the VHF band of the standard German R/T sets and to jam 30–33 MHz (*Ottokar*) and later 38–42 MHz (*Benito*, R/T and *Beam*) wavebands.

On the eve of D-Day in June 1944, 100 Group successfully began its work of deceiving the enemy using the *Mandrel* airborne screen and *Window* feint forces, which gave the illusion on German radar screens that a Main Force was heading for a German target.

The B-24 Liberator was capable of carrying up to as many as 30 jamming sets and this and its long-range capability made it a more acceptable aircraft for the task than the B-17. However, the B-17s played a very important part in 100 Group RCM and ELINT operations, carrying out many different types of jamming using *Jostle*, *Mandrel*, *ABC* and *Airborne Grocer* jammers. The massive *Jostle IV* transmitter unit was housed vertically in the bomb bay of a B-17. No other bomber in British service had a bomb bay deep enough to carry this equipment. Fortresses carried out jamming of German *Düppel (Window)* using recorded engine noise, and voice transmissions by German-speaking Special Wireless Operators broadcast false and misleading information to German nightfighters. Eighty-five B-17Gs went to the RAF as the Fortress III where they were used by Coastal Command and in 1944–45 by two radar countermeasure squadrons.

By the time fighting in Europe ended, almost 300,000 Fortress missions had been launched from British airfields. Altogether, a total of 8,680 B-17Gs were built. Boeing built 4,035 B-17Gs and Douglas and Lockheed built 2,395 and 2,250, respectively. At the peak of B-17 production in June 1944, the Boeing Seattle factory was rolling out 16 Fortresses every 24 hrs. The last B-17G rolled off the Lockheed Vega production line on 29 July 1945.

Thirty-calibre machine-guns could be mounted in sockets in the nose of the B-17E and B-17F for operation by the bombardier and navigator. These sockets, including one non-standard fitting in the roof, can be seen in this B-17E, pictured in the USA. Note also the early ring-and-bead sight atop the Browning, and the empty side magazine. *(USAF)*

The waist-gun positions on the B-17F were cramped and confined, so the staggered waist-gun positions with K-5 mounts and permanently closed one-piece, flush-fitting windows of late 'G' models were much appreciated. The two machine-guns were also installed on the lower sill on a K-6 mount instead of on the older K-5 swivel post. *(USAF)*

Weaponry/Countermeasures

Browning .50 in M2 machine-gun
The standard machine-gun on all US fighters and bombers was the .50 in Browning M2. First produced in 1921, this was basically an enlarged version of the .30 in M1917A1. It was produced in greater numbers than any other US machine-gun in the Second World War:

Chaff (also known as *Window*)
This consisted of metallic foil strips, of varying lengths depending on the radar to be jammed, dropped in huge quantities to reflect radar signals and give enemy operators false readings. The RAF first used it in July 1943, by the 8th AF in December 1943 and the 15th AF in March 1944.

AN/APS-15 (*H2X* or *Mickey*)
Usually known as *H2X*, this was a US-developed version of the British ground-mapping *H2S* radar system. The first use of *H2X* was made on the mission to Wilhelmshaven in October 1943. Sets were available in sufficient numbers by late 1943 to mount the first

major raids to test its effectiveness, and it became the standard device for bombing through overcast conditions. Using *H2X* on D-Day, 1,365 8th AF bombers dropped 2,798 tons of bombs through cloud behind the beachheads thirty minutes before the landing.

AN/APQ-9 and AN/APR 4 (*Carpet III*)
The AN/APQ-9 (transmitter) and AN/APR 4 (search receiver) were used to jam signals operating on the 300–1,000 MHz range (the frequencies of German fire-control radars). The 8th AF in October 1943 and the 15th AF first used the system in August 1944. A 1944 study indicated that formations protected by *Carpet* averaged 1.5 per cent losses, compared to 14.1 per cent losses in unprotected formations. Three-quarters of the bombers in Europe were fitted with *Carpet* systems by the end of the war.

AN/APT-2 and AN/APT-5
These transmitters for spot-jamming German early warning radar transmissions were fitted extensively to bombers in the ETO during the last twelve months of the war in Europe.

Appendix 1
Weapons

Field modifications to B-17E 41-2432 *The Last Straw* of the 63rd Bomb Squadron, 43rd Bomb Group, showing an early attempt to improve the forward firepower of the Fortress in the Pacific by installing tail-guns in the nose. *(USAF)*

Tail-gun station on a B-17E pictured on 1 October 1941 showing the early style ring-and-bead sight mounted outside the window, which was replaced on late model Fortresses two years later by an internal reflector sight. *(Boeing)*

B-17E waist gunners demonstrate how they would use their hand-operated, K-5 post-mounted .50 calibre machine-guns in combat. The metal ammunition boxes (note the two spare) each contain 100 rounds, but these were replaced later by belt feeds with two ammunition boxes being fixed to the roof. *(Boeing)*

B-17E waist gunner keeps watch during a mission over New Guinea. The mount allowed him to swivel his machine-gun inside and outside the opening. Later, all .50s were power-operated, armour plate was installed, and the waist positions were staggered to ease congestion in the compartment. *(USAF)*

B-17 bombardier's station in the glazed nose of a Flying Fortress. On the left hand side is the bomb selecting mechanism. *(via Tom Cushing)*

Waist and ball turret gun installations on Fortress III/B-17G-50-BO 42-102439 of the RAF. Note that the waist window has been glazed over (early models had open hatches) and the gun is fired through a mounting set in the glass. *(Boeing)*

The biggest threat to American bombers over Europe was from *Luftwaffe* fighters making head-on attacks. Here, a B-17 gunner fires a .50-in machine-gun in the nose of a Fortress. *(via Robert M. Foose)*

Early models of the B-17G had the conventional tail-gun installation as used in the 'F' series. The United Air Lines modification centre at Cheyenne, Wyoming, developed a modified tail-gun arrangement, to give the twin fifties greatly increased elevation and azimuth, while the old ring-and-bead sight outside the window was replaced with a new N-8 reflector sight inside the window. The *Cheyenne* or *Pumpkin* installation (which reduced the B-17's overall length by 5 in) also improved all-round visibility by the fitting of larger windows. It was used in the B-17G-80-BO (from 43-38473), -50-DL (from 44-6251), -55-VE (from 44-8287) and subsequent batches. *(via Robert M. Foose)*

Appendix 2
B-17 Specifications

Model	XB-17 (299)	Y1B-17 (299B)	Y1B-17A (299F)	B-17B 299M)	B-17C (299H)	B-17D (299-H)	B-17E (299-0)	B-17F (299-0)	B17G (299-0)
Crew	8	6	6	6	9	10	6–9	10	10
Span	103 ft 9in (for all models)								
Length	68ft 9in	68ft 4in	68ft 4in	67ft 9in	67ft 9in	67ft 11in	73ft 10in	74ft 9in	74ft 4in
Height	14ft 11in	18ft 4in	18ft 4in	18ft 4in	18ft 4in	18ft 4in	19ft 2in	19ft 1in	19ft 1in
Wing Area	1,420 sq ft (for all models)								
Weights (lb)									
Empty	21,657	24,465	31,160	27,650	30,600	30,960	32,250	34,000	36,135
Max	38,053	39,000	45,650	46,178	49,650	49,650	54,000	65,500	65,500
Engines Four each of	750hp P&W Hornet S1EG	930hp Wright Cyclone R-1820-39	1,000hp Wright Cyclone R-1820-51	1,200hp Wright Cyclone R-1820-51	1,000hp Wright Cyclone R-1820-65	1,200hp Wright Cyclone R-1820-65	1,200hp Wright Cyclone R-1820-65	1,380hp Wright Cyclone R-1820-97	1,380hp Wright Cyclone R-1820-98
Performance									
Max speed	236mph at 14,000ft	256mph at 25,000ft	271mph at 25,000ft	286mph at 25,000ft	323mph at 25,000ft	323mph at 25,000ft	317mph at 25,000ft	325mph at 25,000ft	302mph at 25,000ft
Cruising	140mph	175mph	183mph	225mph	227mph	227mph	195mph	160mph	160mph
Climb rate	8 m to 10,000ft	6m to 10,000ft	7m 48s 10,000ft	7m to 10,000ft	7m to 10,000ft	7m 12s to 10,000ft	7m 6s to 20,000ft	25m 42s 20,000ft	37m to 20,000ft
Ceiling	24,620ft	30,600ft	38,000ft	30,000ft	37,000ft	37,000ft	36,000ft	37,500ft	35,600ft
Range (ferry)	3,011	2,400– 3,400	2,400– 3,600	2,400– 3,600	2,000– 3,400	2,000– 3,400	2,000– 3,200	1,300– 3,800	1,140 miles 2,740 miles
Machine-guns	5x.30	5x.30 or .50	5x30 or .50	1x.30 6x.50	1x.30 6x.50	1x.30 6x.50	1x.30 8x.50	11x.50	12/13x.50
Bomb load	8x600-lb (max)	8x600-lb (max)	8x600-lb (max)	4x1100-lb or 20x100-lb	8x600-lb or 4x1100-lb or 20x100-lb	8x600-lb or 4x1100-lb or 20x100-lb	4x1000-lb or 20x100-lb or 14x300-lb	8x1000-lb or any combination to 24x100-lb	6x1600-lb and 2x4000-lb

Appendix 3
Fortress Production

TOTAL PRODUCTION FIGURES, BOEING B-17

TOTAL PRODUCTION FIGURES, BOEING B-17	
EXPERIMENTAL MODELS	15
B-17B	39
B-17C	38
B-17D	42
B-17E	512
B-17F	3,400
B-17G	8,680
Total	**12,726**

B-17G Production

Manufactured by Boeing, Seattle WA		Manufactured by the Douglas Aircraft Co., Long Beach CA		Manufactured by the Vega Aircraft Corp., Burbank CA	
SERIAL No.	QTY	SERIAL No.	QTY	SERIAL No.	QTY
42-31032/42-32116	1,085	42-23563	1	42-39758/42-40057	300
42-97058/42-97407	350	42-37716	1	42-97436/42-98035	600
42-102379/42-102978	600	42-37721/42-38213	493	44-8001/44-9000	1,000
43-37509/43-39508	2,000	42-106984/42-107233	250	44-85492/44-85841	350
		44-6001/44-7000	1,000		
		44-83236/44-83885	650		
Total	**4,035**	**Total**	**2,395**	**Total**	**2,250**

Model	Serial No	No.	Model	Serial No	No.	Model	Serial No	No.
Model 229	(NX-13372)	1	B-17F-20-VE	42-5765/42-5804	40	B-17G-45-DL	44-6126/44-6250	125
Y1B-17A	36-149/36-161	13	B-17F-25-VE	42-5805/42-5854	50	B-17G-50-DL	44-6251/44-6500	250
Y1B-17A	37-369	1	B-17F-30-VE	42-5855/42-5904	50	B-17G-55-DL	44-6501/44-6625	125
B-17B	38-211/38-220	10	B-17F-35-VE	42-5905/42-5954	50	B-17G-60-DL	44-6626/44-6750	125
B-17B	38-221/38-223	3	B-17F-40-VE	42-5955/42-6029	75	B-17G-65-DL	44-6751/44-6875	125
B-17B	38-258/38-270	13	B-17F-45-VE	42-6030/42-6104	75	B-17G-35-BO	42-31932/42-32116	185
B-17B	38-583/38-584	2	B-17F-50-VE	42-6105/42-6204	100	B-17F-80-DL	42-37714/42-37715	125
B-17B	38-610	1	B-17F-55-BO	42-29467/42-29531	65	B-17G-10-DL	42-37716	1
RB-17B	39-1/39-10	10	B-17F-60-BO	42-29532/42-29631	100	B-17F-80-DL	42-37717/42-37720	125
RB-17C	40-2042/40-2079	38	B-17F-65-BO	42-29632/42-29731	100	B-17G-10-DL	42-37721/42-37803	83
RB-17D	40-3059/40-3100	42	B-17F-70-BO	42-29732/42-29831	100	B-17G-15-DL	42-37804/42-37893	90
B-17E	41-2393/41-2669	277	B-17F-75-BO	42-29832/42-29931	100	B-17G-20-DL	42-37894/42-37988	95
B-17E	41-9011/41-9245	235	B-17F-80-BO	42-29932/42-30031	100	B-17G-25-DL	42-37989/42-38083	95
B-17F-1-BO	41-24340/41-24389	50	B-17F-85-BO	42-30032/42-30131	100	B-17G-30-DL	42-38084/42-38213	130
B-17F-5-BO	41-24390/41-24439	50	B-17F-90-BO	42-30132/42-30231	100	B-17G-1-VE	42-39758/42-39857	100
B-17F-10-BO	41-24440/41-24489	50	B-17F-95-BO	42-30232/42-30331	100	B-17G-5-VE	42-39858/42-39957	100
B-17F-15-BO	41-24490/41-24503	14	B-17F-100-BO	42-30332/42-30431	100	B-17G-10-VE	42-39958/42-40057	100
B-17F-20-BO	41-24504/41-24539	36	B-17F-105-BO	42-30432/42-30531	100	B-17G-40-BO	42-97058/42-97172	115
B-17F-25-BO	41-24540/41-24584	45	B-17F-110-BO	42-30532/42-30616	85	B-17G-45-BO	42-97173/42-97407	235
B-17F-27-BO	41-24585/41-24639	55	B-17F-115-BO	42-30617/42-30731	115	B-17G-15-VE	42-97436/42-97535	100
B-17F-1-DL	42-2964/42-2966	3	B-17F-120-BO	42-30732/42-30831	100	B-17G-20-VE	42-97536/42-97635	100
B-17F-5-DL	42-2967/42-2978	12	B-17F-125-BO	42-30832/42-30931	100	B-17G-25-VE	42-97636/42-97735	100
B-17F-10-DL	42-2979/42-3003	25	B-17F-130-BO	42-30932/42-31031	100	B-17G-30-VE	42-97736/42-97835	100
B-17F-15-DL	42-3004/42-3038	35	B-17G-1-BO	42-31032/42-31131	100	B-17G-35-VE	42-97836/42-97935	100
B-17F-20-DL	42-3039/42-3073	35	B-17G-5-BO	42-31132/42-31231	100	B-17G-70-DL	44-6876/44-7000	125
B-17F-25-DL	42-3074/42-3148	75	B-17G-10-BO	42-31232/42-31331	100	B-17G-45-VE	44-8001/44-8100	100
B-17F-30-DL	42-3149/42-3188	40	B-17G-15-BO	42-31332/42-31431	100	B-17G-50-VE	44-8101/44-8200	100
B-17F-35-DL	42-3189/42-3228	40	B-17G-20-BO	42-31432/42-31631	200	B-17G-55-VE	44-8201/44-8300	100
B-17F-40-DL	42-3229/42-3283	55	B-17G-25-BO	42-31632/42-31731	100	B-17G-60-VE	44-8301/44-8400	100
B-17F-45-DL	42-3284/42-3338	55	B-17G-30-BO	42-31732/42-31931	200	B-17G-65-VE	44-8401/44-8500	100
B-17F-50-DL	42-3339/42-3393	55	B-17G-40-VE	42-97936/42-98035	100	B-17G-70-VE	44-8501/44-8600	100
B-17F-55-DL	42-3394/42-3422	55	B-17G-50-BO	42-102379/42-102543	165	B-17G-75-VE	44-8601/44-8700	100
B-17F-60-DL	42-3423/42-3448	26	B-17G-55-BO	42-102544/42-102743	200	B-17G-80-VE	44-8701/44-8800	100
B-17F-65-DL	42-3449/42-3482	34	B-17G-60-BO	42-102744/42-102978	235	B-17G-85-VE	44-8801/44-8900	100
B-17F-70-DL	42-3483/42-3503	21	B-17G-35-DL	42-106984/42-107233	250	B-17G-90-VE	44-8901/44-9000	100
B-17F-75-DL	42-3504/42-3562	59	B-17G-65-BO	43-37509/43-37673	165	B-17G-75-DL	44-83236/44-83360	125
B-17G-5-DL	42-3563	1	B-17G-70-BO	43-37674/43-37873	200	B-17G-80-DL	44-83361/44-83485	125
B-17F-30-BO	42-5050/42-5078	29	B-17G-75-BO	43-37874/43-38073	200	B-17G-85-DL	44-83486/44-83585	100
B-17F-35-BO	42-5079/42-5149	71	B-17G-80-BO	43-38074/43-38273	200	B-17G-90-DL	44-83586/44-83685	100
B-17F-40-BO	42-5150/42-5249	100	B-17G-85-BO	43-38274/43-38473	200	B-17G-95-DL	44-83686/44-83885	200
B-17F-45-BO	42-5250/42-5349	100	B-17G-90-BO	43-38474/43-38673	200	B-17G-95-VE	44-85492/44-85591	100
B-17F-50-BO	42-5350/42-5484	135	B-17G-95-BO	43-38674/43-38873	200	B-17G-100-VE	44-85592/44-85691	100
B-17F-1-VE	42-5705/42-5709	5	B-17G-100-BO	43-38874/43-39073	200	B-17G-105-VE	44-85692/44-85791	100
B-17F-5-VE	42-5710/42-5724	15	B-17G-105-BO	43-39074/43-39273	200	B-17G-110-VE	44-85792/44-85841	50
B-17F-10-VE	42-5725/42-5744	20	B-17G-110-BO	43-39274/43-39508	200			
B-17F-15-VE	42-5745/42-5764	20	B-17G-40-DL	44-6001/44-6125	125			

Appendix 4
Postwar Civil B-17s and their Fates

Type/Serial No.	Registration	Last Owner	History
B-17E 41-9210	CP-753	Frigorificos Reyes, Bolivia	Crash-landing San Borja 8.76. Being rebuilt at La Paz
B-17F 41-24434	CP-579	Frigorificos Grigota	Crashed Uncio 29.12.58
B-17F 42-3217		Swedish Government	*Georgia Rebel,* 381st BG, landed at Vannacka, Sweden 24.7.43 and interned. Bought by ABA. Fate unknown
B-17F 42-3360	CP-570	Lloyd Aero Boliviano	Crashed La Paz 21.9.55
B-17F 42-3470	HK-580	Linea Aerea Borinquen	Converted to TB-17F 12.53. Ultimate fate unknown
B-17F 42-3490	SE-BAN	SILA (Swedish International Airline)	385th BG. Landed at Bulltofta, Sweden, and interned on 21.6.44. Converted by SAAB to 14-passenger airliner and bought by ABA as *Bob* (10.9.45 5.10.48) Broken up 5.10.48
B-17F 42-3543	SE-BAH	ABA	*Sack Time Suzy,* 96th BG, landed at Bulltofta and interned 9.10.43. Converted by SAAB to 14-passenger airliner *Sam* in 1945. Used for fire practice 9.46
B-17F 42-5827		Swedish Government	*Lakanuki,* 379th BG, landed at Sattarp, Sweden, 1.4.44 and interned. Bought by ABA. Fate unknown
B-17F 42-6073	CP-686	Lloyd Aero Boliviano	Converted to TB-17F. Crashed Trinidad, Bolivia, in the 1960s
B-17F 42-6107	N1340N	Aero Flite Inc, Cody, WY	Converted to TB-17F. Re-engined with 4 RR Darts ex-UAL Viscount after 1968. Crashed Dubois, WY, while fire-bombing on 18.8.70
B-17F 42-30177	F- BGSG	Institute Geographique Nationale	Served French Air Force as *Charlene, Bir Hakiem,* based Wahn, FGR, for French High Commissioner. Damaged and ferried to IGN, Creil, France Used for spares
B-17F 42-30661	SE-BAK	ABA	388th BG *Veni Vidi Vici* landed Rinkaby, Sweden and interned 29.2.44. Was on its 28th mission. Converted by SAAB to 14-passenger airliner *Jim.* Broken up 12.46
B-17F 42-30921	CP-571	Aerovias Moxos	Lloyd Aero Boliviano in 50s. Crashed Viacha 16.1.62
B-17F 42-31163	SE-BAM	ABA	*A Good Ship & A Happy Ship* 388th BG, landed Rinkaby, Sweden and interned 6.3.44. Converted by SAAB to 14-passenger airline *Tom.* Crashed Mariefred 4.12.45
B-17G 42-97115	SE-BAO	SILA	94th BG/333rd BS. Landed Bulltofta, Sweden and interned 11.4.44. Converted by SAAB to 14-passenger airliner *Ted.* Broken up and struck off register 5.10.48
B-17G 42-102542	N5845N	Aero Service Corp. Philadelphia, PA	Converted to TB-17G. Served as memorial at Sioux City municipal airport *Sioux City Sue* 1946-52. Crashed c1958
B-17G 42-102715	N66573	Black Hills Aviation, Alamogordo, NM	Converted to TB-17G. Fire tanker 20.11.61-77. Crashed
B-17G 42-107067	OY-DFE	DDL (Denmark)	447th BG/709th BS. Diverted to Sweden and interned 11.4.44. Converted by SAAB to 14-passenger airliner 19.11.45. Written off in collision Collided with aircraft landing at Kastrup 30.1.46.
B-17G 43-37650	N66570	California Atlantic Airways, Florida	Formerly with 96th BG. Fate unknown after 1952
B-17G 43-38322	CP-936	Frigorificos Reyes (1971-)	Oklahoma Military Academy, Rogers City, OK, 1946. Crashed La Paz, Bolivia 7.2.65 while serving with Lloyd Aero Boliviano. Rebuilt. Crashed San Ignacio de Moxos 11.2.72
B-17G 43-38978	N4960	Columbia Airmotive, Troutdale, OR	Dismantled for spares in 1950s
B-17G 43-39304	F-BDAT	Institute Geographique Nationale	Crashed Niamey, Nigeria 12.12.50
B-17G 43-39367	CP-625	Lloyd Aero Boliviano (11.56-)	Crashed San Lorcnzo, Bolivia 17.11.59
B-17G 44-6332	CP-588	Aerovias Moxos	Crashed 2.5.63
B-17G 44-6556	CP-624	Corporation Boliviano de Fomento	Crashed Reyes, Bolivia 23.2.63
B-17G 44-8990	N3678G	Multiple Management Corp. Long Beach	Fate unknown after 1969
B-17G 44-6750	CP-597	Lloyd Aero Boliviano (11.52-)	Crashed Trinidad, Bolivia 5.9.55
B-17G 44-83439	N131P	Air Carrier Leasing Corp.	Used by Paramount Aquariums Inc, Veto Beach,Miami, FL (1963-). Converted for fish transport. Fate unknown
B-17G 44-83728	F- BGOE	Institute Geographique Nationale	Based at Creil, France from 12.7.52. Fate unknown

Type/Serial No	Registration	Last Owner	History
B-17G 44-83729	F-BEED	Institute Geographique Nationale	*Denise*. Struck off register at Creil 1962 and scrapped
B-17G 44-83750	CP-623	Lloyd Aero Boliviano (11.56-)	Crashed La Paz, Bolivia 28.7.58
B-17G 44-83753	NL5024N	A. Schwimmer Miami/Israeli Air Force (9.6.48)	Scrapped in Israel c1958
B-17G 44-83757	F-BDRR	Institute Geographique Nationale	Scrapped at Creil c1972
B-17G 44-83809	CP-626	Corporation Boliviano de Fomento	Crashed Caranavi 25.10.59
B-17G 44-83811	N9814F	Israeli Air Force (9.6.48-58)	Impounded in Algeria en route to USA 1958. Returned to Israel. Dismantled. Fuselage to Columbia Pictures Corp. UK for use as prop in movie *The War Lover* 1961
B-17G 44-83842	NL1212N	Charles Babb Inc. Burbank, CA	A. Schwimmer, Miami/Israeli Air Force 6.48. Impounded Azores on ferry to Israel. Used by Portuguese AF, based Azores. Fate unknown
B-17G 44-8385 1	NLI098M	Charles T. Winters, Miami, FL	A. Schwimmer/Israeli Air Force 9.6.48. Scrapped c.1962
B-17G 44-83858	N7228C	Aero Enterprises, Fresno, CA (1966)	Bu77226 (PB-1W). Fate unknown
B-17G 44-83858	CP-742	Comp. Boliviana de Rutes Aereas (64)	Bu77227 (PB-1W). Crashed Santa Ana 21.2.65
B-17G 44-838 59	CP-767	Comp. Boliviana de Rutes Aereas (65)	Bu77228 (PB-1W). Last USN PB-1W, disposed of from Litchfield Park NAS, Phoenix 1958. Crashed Santa Ana 13.4.67
B-17G 44-83861	CP-741	Comp. Boliviana de Rutes Aereas (64)	Bu77229 (PB-1W). Crashed San Borla 30.10.64
B-17G 44-83864	N73648	Black Hills Aviation, Spearfish, SD	Bu77232 (PB-1W). Crashed Silver City, NM 12.7.72
B-17G 44-83873	CF-JJH	Unknown	Bu77236 (PB-1W). Fate unknown
B-17G 44-83874	N5236V	American Compressed Steel Corp.	Bu77237 (PB-1W). Based at Love Field, Dallas, TX 66/69. Fate unknown
B-17G 44-83875	CP-640	Boliviana de Aviacion (1958-)	Bu77238 (PB-1W). Crashed La Paz, Bolivia 17.8.67
B-17G 44-83877	N5232V	Columbia Pictures Inc (1961-)	Bu77240 (PB-1W). Ferried to UK for *The War Lover*. Arrived Gatwick 8.10.61. Scrapped at Bovingdon 1962
B-17G 44-83883	N5229V	Columbia Pictures Inc (1961-)	Bu77243 (PB-1W). Ferried to UK for movie *The War Lover*. Arrived Gatwick 8.10.61. Scrapped at Manston 1962
B-17G 44-85507	N5116N	Mark Hurd Mapping Co, Minneapolis	Assigned to Esperado Mining Co, Altus, OK for scrapping 25.6.47. Converted to exec aircraft for *Chicago Tribune* 4.2.48. Crashed in Nevada desert 10.11.52
B-17G 44-83861	CP-741	Comp. Boliviana de Rutes Aereas (64)	Bu77229 (PB-1W). Crashed San Borja 30.10.64
B-17G 44-83864	N73648	Black Hills Aviation, Spearfish, SD	Bu77232 (PB-1W). Crashed Silver City, NM 12.7.72
B-17G 44-83873	CF-JJH	Unknown	Bu77236 (PB-1W). Fate unknown
B-17G 44-83874	N5236V	American Compressed Steel Corp.	Bu77237 (PB-1W). Based at Love Field, Dallas, TX 66/69. Fate unknown
B-17G 44-83875	CP-640	Boliviana de Aviacion (1958-)	Bu77238 (PB-1W). Crashed at La Paz, Bolivia 17.8 .67
B-17G 44-83877	N5232V	Columbia Pictures Inc (1961-)	Bu77240 (PB-1W). Ferried to UK for *The War Lover*. Arrived Gatwick 8.10.61. Scrapped at Bovingdon 1962
B-17G 44-83883	N5229V	Columbia Pictures Inc (1961-)	Bu77243 (PB-1W). Ferried to UK for *The War Lover*
B-17G 44-85594	F- BGSQ	Institute Geographique Nationale	Scrapped at Creil c1973
B-17G 44-85643	F-BEEA	Institute Geographique Nationale	*Amelie*, Chateau de Verneil. Acquired for filming of *Memphis Belle*. Destroyed in crash at Binbrook 25.7.89
B-17G 44-85728	F- BGOE	Institute Geographique Nationale	Previous owners included TWA, Kansas City, MO 26.6.46. Exec conversion by Boeing as special Model 299AB. Acquired by Shah of Persia 4.47 before being moved to France. Last flight with IGN 22.8.67. Scrapped 1970
B-17G 44-85733	F-BEEB	Institute Geographique Nationale	Crashed Yaounde, Cameroon 11.3.49
B-17G 44-85774	N621L	Aircraft Specialties Inc, Mesa, AZ	Lloyd Aero Boliviano 12.56-68. Fire bomber 68-72. Crashed 7.75
B-17G 44-85778	N3509G	Aircraft Component Equipment Supplies	Klamath Falls, OR, Fire bomber with various companies 1967-77. Fate unknown
B-17G 44-85806	CP-762	Bolivian Air System (12.64-)	Crashed at La Paz 16.12.64
B-17G 44-85812	N4710C	Dothan Aviation Carp, Dothan, AL	Bu77246 (PB-1G USCG). Crashed Blakely, GA 5.8.76
B-17G 44-858 13	N6694C	Black Hills Aviation, Alamogordo, NM	EB-17G. JB-17G converted to Model 299Z by Boeing, Wichita. Converted as engine testbed, 5th position in nose. Sold 1981. Struck off USCR by 1984
B-17G 44-85817	CP-622	Aerovias Los Andes (10.56-)	Crashed Laja 18.2.57
B-17G 44-85821	N2873G	Unknown	Bu77247 (PB-1G USCG) until 1959. Fate unknown
B-17G 44-85824	CP-694	Servicios Aereos Cochabamba (1960-)	Bu77250 (PB-1G USCG). Crashed La Paz, 18. 12.63
B-17G 44-85840	N620L	Aircraft Specialties Inc, Mesa, AZ	Used by Lloyd Aero Boliviano (CP-620) 11.56. Took part in *Tora! Tora! Tora!* in Hawaii 1969. Scrapped c1974
B-17G 44-?	N5225V	American Compressed Steel Corp, TX	Bu77138 (PB-1W) 1963. Fate unknown
B-17F ?	N60475	Charles Babb & Co. New York	Formerly used by Sperry Gyroscope Co. (4.1.52). Fate unknown

Appendix 5
Museum Aircraft

Model	Serial	Name	Location
B-17D (RB-17D)	40-3097	*Swoose*	Paul F. Garber Restoration Facility, Silver Hill, Maryland
B-17E	41-2446	*Swamp Ghost*	Agaiambo Swamp, Papua, New Guinea
B-17E (XC108A)	41-2595		In storage at Gait Airport, Illinois
B-17E	41-9101		Abandoned in Greenland. Now under 260 ft of ice
B-17E	41-9105		Abandoned in Greenland. Now under 260 ft of ice
B-17E	41-9210/N8WJ		Scott D. Smith, Colorado Springs, CO
B-17F-10-BO	41-24485	*Memphis Belle*	Memphis Belle Memorial Association, Memphis, Tennessee
B-17F-50-BO	42-3374		In storage at Offutt AFB, Nebraska
B-17F-70-BO	42-29782/N17W		Museum of Flight, Seattle, Washington
B-17G-35-BO	42-32076	*Shoo Shoo Baby*	Wright-Patterson AF Museum, Dayton, Ohio
B-17G-90-BO	43-38635/N3702G	*Virgin's Delight*	Castle Air Museum, Castle AFB, Merced, California
B-17G-50-DL	44-6393/CP-891	*2nd Patches*	March Field Museum, March AFB, CA. (As 42-30092)
B-17G-70-VE	44-8543/N3701G	*Chuckie**	BC Vintage Flying Machines, Fort Worth, Texas
B-17G-85-VE	44-8846/F-AZDX	*Pink Lady**	Association Fortress Volante. Jean Salis, Cerny, France
B-17G-85-VE	44-8889/F-		*Musée de l'Air*, Le Bourget, France
B-17G-75-DL	44-83316		Stored in pieces at Ocotillo Wells, California
B-17G-85-DL	44-83512	*Heavens Above*	Lackland AFB History & Tradition Museum, San Antonio, TX
B-17G-85-DL	44-83514/N9323Z	*Sentimental Journey**	Arizona Wing of the Confederate Air Force, Mesa, Arizona
B-17G-85-DL	44-83525/N83525	*Suzy Q*	Stored in pieces Fantasy of Flight, Polk County, Florida
B-17G-85-DL	44-83542/N9324Z	*Piccadilly Princess*	Fantasy of Flight, Weeks Coll
B-17G-85-DL	44-83546/N3703G	*Memphis Belle*	March AFB, California
B-17G-85-DL	44-83559	*King Bee*	Strategic Air Command Museum, Offutt AFB, Nebraska
B-17G-85-DL	44-83563/N9563Z	*Fuddy Duddy**	National Warplanes Museum, New York
B-17G-85-DL	44-83575/N93012	*909**	Collings Foundation, Riverhill Farm, Stow, Maine
B-17G-90-DL	44-83624		Dismantled for restoration, Dover AFB, Delaware
B-17G-90-DL	44-83663/N47780	*Short Bier*	Hill AFB Museum, Utah
B-17G-90-DL	44-83684/N3713G	*Piccadilly Lilly*	Planes of Fame Museum, Corona del Mar, California
B-17G-95-DL	44-83690	*Miss Liberty Belle*	Grissom AFB Museum Foundation, Peru, Indiana
B-17H-95-DL	44-83718		*Museu Aerospacial*, Rio de Janeiro, Brazil (also 44-83462)
B-17G-95-DL	44-83722		Stored in pieces at Ocotillo Wells, California
B-17G-95-DL	44-83728		*Musée de l'Air*, Le Bourget, France
B-17G-90-DL	44-83735/F-BDRS	*Mary Alice*	Imperial War Museum, Duxford, England
B-17G-95-DL	44-83785/N207EV	*Shady Lady**	Evergreen Helicopters, Pinal Airpark, Marana, Arizona
B-17G-95-DL	44-83790		Abandoned in Newfoundland 1947. Loc 1970 almost intact
B-17G-95-DL	44-83814/N66571	*Tanker 09*	In storage at Dulles Airport, near Washington DC
B-17G-95-DL	44-83863/N5233V		USAF Armament Museum, Eglin AFB, Eglin, Florida
B-17G-95-DL	44-83868/N5237V		RAF Bomber Command Museum, Hendon, London
B-17G-95-DL	44-83872/N7227C	*Texas Raiders**	Gulf Coast Wing, Confederate Air Force, Harlingen, Texas
B-17G-95-DL	44-83884/N5230V	*Yankee Doodle II*	8th Air Force Museum, Barksdale AFB, Bossier City, LA
B-17G-95-DL	44-85583		On display at base area, de Recife, Brazil
B-17G-100-VE	44-85599	*Blackhawk*	Texas Museum of Military History, Dyess AFB, Abilene, Texas
B-17G-105-VE	44-85718/N900RW	*Thunder Bird**	Lone Star Museum, Hobby Airport, Houston, Texas
B-17G-105-VE	44-85734/N5111N	*Five Engine*	New England Air Museum, Bradley Airport, Windsor Locks CT
B-17G-105-VE	44-85738	*Preston'sPride*	American Veterans Memorial, Tulare, CA (as 0-85738)
B-17G-105-VE	44-85740/N5017N	*Aluminum Overcast**	EAA Warbirds of America, Oshkosh, Wisconsin
B-17G-105-VE	44-85778/N3509G		Semi-storage, Stockton, California
B-17G-105-VE	44-85784/G-BEDF	*Sally B**	B-17 Preservation Ltd, a/c based at Duxford
B-17G-105-VE	44-85790		On top of Bomber Gas Station, Milwaukee, Oregon
B-17G-110-VE	44-85813/N6694C		Stored at Kissimmee, Florida
B-17G-110-VE	44-85825		Smithsonian Institute, Washington DC
B-17G-110-VE	44-85828/N9323R	*I'll Be Around*	390th BG Assn, Pima Air Museum, Tucson, Arizona
B-17G-110-VE	44-85829/N31 930	*Yankee Lady*	Yankee Air Force, Ypsilanti, Missouri

(BO) Boeing (DL) Doug as (VE) Lockheed Vega   *Flying Examples

Appendix 6
Fortress Models, Decals and Parts

Manufacturer	Scale	Part no.	Item
ACADEMY		AC1666	Boeing B-17C Fortress
ACADEMY		AC1683	Boeing B-17D
ACADEMY		AC2102	Boeing B-17F Fortress
ACADEMY	1:24	AC2106	Boeing B-17B Fortress
ACADEMY	1:72	AC2141	Boeing B-17E RAF Coastal Command
ACADEMY	1:72	AC2142	Boeing B-17F *Fort Alamo*
ACADEMY	1:72	AC2143	Boeing B-17G Nose Art
ACADEMY	1:72	AC2150	Boeing B-17C/D *Kelly*
ACADEMY	1:72	AC2165	Boeing SB-17 with fuselage
ACADEMY	1:72	AC2188	Boeing B-17F *Memphis Belle*
ACADEMY	1:72	AX05005	B-17G Flying Fortress
AIRFIX	1:72	AX08005	Boeing B-17G
AIRFIX	1:72	HA00068	B-17G 381st Bomb Group
HASEGAWA	1:72	HA00077	Boeing B-17F Flying Fortress
HASEGAWA	1:72	HAK010	B-17G Flying Fortress
HASEGAWA	1:72	HAK010X	B-17G Flying Fortress
HASEGAWA	1:72	HAK011	B-17F
HASEGAWA	1:72	HAK011X	Boeing B-17F Flying Fortress
HASEGAWA	1:72	HAK106	Boeing B-17G *Super Rabbit*
HASEGAWA	1:72	HAK109	Boeing B-17F *My Devotion*
HASEGAWA	1:72	HASP018	B-17F with extra nose art
HASEGAWA	1:72	HASP023	B-17G with extra nose art
HASEGAWA	1:72	HASP117	B-17G Flying Fortress Nose
HASEGAWA	1:72	ID32B1	B-17 Flying Fortress
ID MODELS Vac Form	1:32	MC14401	Boeing B-17G Fortress
MINICRAFT	1:14	4 MG5600	B-17G Flying Fortress
MONOGRAM	1:48	MG5928	B-17G with Cheyenne tail gun
MONOGRAM	1:48	PK603	B-17G Flying Fortress
MATCHBOX	1:72	RV4395	B-17F Fortress
REVELL	1:72	RV4569	Boeing B-17G Flying Fortress
REVELL	1:48	RV4701	B-17F *Memphis Belle*
AEROMASTER DECAL		AMD48415	B-17 Fortresses in the Sky 1
AEROMASTER DECAL	1:48	AMD48416	B-17 Fortresses in the Sky 2
AEROMASTER DECAL	1:72	AMD72174	B-17 Fortress
CARPENA DECALS	1:48	CA48031	B-17 Flying Fortress
CARPENA DECALS	1:72	CA72058	B-17 Flying Fortress
SKYMODELS DECALS	1:72	SKY72007	B-17F/G
SUPERSCALE DECAL	1:72	SS48017	B-17G (4) 323BS OR-R; OR-N; J
SUPERSCALE DECAL	1:48	SS48021	B-17F/G (3) 322BS; 463rd Bomb Group; 447
SUPERSCALE DECAL	1:48	SS48059	B-17F/G (4) 91st Bomb Group; *8 Ball* OR-H
SUPERSCALE DECAL	1:48	SS48320	B-17F
SUPERSCALE DECAL	1:48	SS48481	B-17G *Mistreated, Good Deal*
SUPERSCALE DECAL	1:48	SS48524	B-17F *Yardbird II, Marnita,*
SUPERSCALE DECAL	1:48	SS48525	B-17G *Shady Lady, False*
SUPERSCALE DECAL	1:48	SS48582	B-17G *£5 with Breakfast*
SUPERSCALE DECAL	1:48	SS48583	B-17G *Carolina Moon*
SUPERSCALE DECAL	1:48	SS48584	B-17G *Looky Looky*
SUPERSCALE DECAL	1:48	SS48585	B-17G *American Beauty*
SUPERSCALE DECAL	1:48	SS48672	B-17F 510BS/ 322BS
SUPERSCALE DECAL	1:48	SS48673	B-17G 324BS/ 401BS 91st Bomb Group 4
SUPERSCALE DECAL	1:48	SS72023	Famous B-17E/F Fortress (5)
SUPERSCALE DECAL	1:72	SS72182	B-17G & YB-17 (4)
SUPERSCALE DECAL	1:72	SS72281	Bomb Groups. B-17 386th
SUPERSCALE DECAL	1:72	SS72547	B-17F (3) 322 BS 91st Bomb Group; 510 BS
SUPERSCALE DECAL	1:72	SS72680	B-17G *Mistreated, Good Deal*
SUPERSCALE DECAL	1:72	SS72713	B-17F/G *Yardbird II*
SUPERSCALE DECAL	1:72	SS72746	B-17G 834 BS; 850 BS
SUPERSCALE DECAL	1:72	SS72747	B-17G 851 BS; 349 BS
SUPERSCALE DECAL	1:72	SS72748	B-17G 851 BS; 323 BS
SUPERSCALE DECAL	1:72	SS72749	B-17G
SUPERSCALE DECAL	1:72	X02172	B-17 & B-24 in RAF Service
XTRADECALS	1:72		DETAILING ACCESSORIES, Photo-etched and resin
AIRWAVES	1:72	CEC48044	B-17C/D/E/F Radio Compartment

Manufacturer	Scale	Part no.	Item
CUTTING EDGE RESIN	1:48	ED48207	Boeing B-17F/G interior
MG EDUARD	1:48	ED48208	Boeing B-17F/G exterior
MG EDUARD	1:48	ED48209	Boeing B-17F/G armament
MG EDUARD	1:48	ED72170	Boeing B-17E/F/G Bomb Bay
AC EDUARD	1:72	ED72171	Boeing B-17 Undercarriage
AC EDUARD	1:72	ED72172	Boeing B-17E/F Interior
AC EDUARD	1:72	ED72173	Boeing B-17 Flaps
AC EDUARD	1:72	EDXS063	Boeing B-17F/G
AC EDUARD	1:72	FH4811	Modified B-17 etched set
FLIGHTPATH	1:48	FHP4801	3 B-17G Flying Fortress
FLIGHTPATH	1:48	MAG7273	Boeing/Saab B-17
AC MAGNA MODELS	1:72	MAQT7203	Boeing B-17E Flying Fortress
MAQUETTE	1:72	PAR48014	B-17 F/G Prop Governors
PARAGON DESIGNS	1:48	PAR48048	PB-1W (B-17G)
PARAGON DESIGNS	1:48	PAR48049	YB-40 Gunship (B-17F)
PARAGON DESIGNS	1:48	PAR48050	B-17 BTO *Mickey Radome*
PARAGON DESIGNS	1:48	PAR48051	Boeing B-17E PA
PARAGON DESIGNS		PAR48052	Boeing B-17 crew access doors
PARAGON		PAR48093	RAF 100 Group B-17G & B-24J/M
PARAGON DESIGNS	1:48	PAR48103	B-17 Flaps
PARAGON DESIGNS	1:48	PAR48104	B-17 Open Bomb Bay
PARAGON DESIGNS	1:48	PAR48105	B-17G Cheyenne Tail Turret
PARAGON DESIGNS	1:48	PAR48106	B-17H/SB-17G with lifeboat
PARAGON DESIGNS	1:72	PAR72004	PB-1W (B-17G)
PARAGON DESIGNS	1:72	PAR72005	YB-40 Gunship (B-17F)
PARAGON DESIGNS	1:72	PAR72006	B-17 BTO *Mickey Radome*
PARAGON DESIGNS	1:72	PAR72042	RAF 100 Group B-17G & B-24J/M
PARAGON DESIGNS	1:72	PAR72043	B-17 Flaps
PARAGON DESIGNS	1:72	PAR72044	B-17H/SB-17G with lifeboat
SQUADRON SIGNAL	1:72	SQS5516	B-17 Flying Fortress
SQUADRON SIGNAL	1:72	SQS6045	B-17 Flying Fortress
SQUADRON SIGNAL	1:72	SQS9137	B-17F/G Canopy, radio room
SQUADRON SIGNAL	1:72	SQS9138	B-17G Nose cone, astrodome
SQUADRON SIGNAL	1:72	SQS9139	B-17C/D Nose cone, canopy
SQUADRON SIGNAL	1:72	SQS9140	B-17F Nose cone, cheek gun
SQUADRON SIGNAL	1:72	SQS9175	Boeing B-17E/F Flying Fortress
SQUADRON SIGNAL	1:72	SQS9176	Boeing B-17G Flying Fortress
SQUADRON SIGNAL	1:72	SQS9551	B-17G Canopy & Ball turret
SQUADRON SIGNAL	1:48	SQS9552	B-17G Nose cone, upper turret
TRUE DETAILS	1:48	TD48010	B-17E-G Flying Fortress
TRUE DETAILS	1:48	TD48098	Boeing B-17E thru G. 2 main
TRUE DETAILS	1:48	TD48477	B-17G Flying Fortress
TRUE DETAILS	1:72	TD72020	B-17 Flying Fortress
TRUE DETAILS	1:72	TD72044	Boeing B-17E thru G. 2 Main
TRUE DETAILS	1:72	TK48040	B-17 Battle damaged tail
TEKNICS	1:48	TV04	B-17 Flying Fortress
TRAVERS	1:48	USK7201	RAF B-17C
80 AVIATION USK DE	1:72	VL0583	B-17 *Memphis Belle* Crewman
VERLINDEN	120MM	VL0718	
VERLINDEN	1:48		B-17G Update Set
VERLINDEN	1:48	XP4814	B-17F/G Fortress Wing Intake
XTRAPARTS	1:48		

Information kindly supplied by David Hannant.
Hannants, Tel: 00 44 (0)1502 517444 (8 lines) Fax: 00 44 (0)1502 500521
Email: david@hannants.co.uk
Web Site: http://www.hannants.co.uk
Postal Address: H. G. Hannant Ltd,
Harbour Road, Oulton Broad, Lowestoft, Suffolk, NR32 3LZ, England.

Appendix 7
Fortress Books

Andrews, Paul M. & Adams, William H. *Heavy Bombers of the Mighty Eighth.* Eighth Air Force Museum Foundation Project Bits & Pieces, 1995

Andrews, Paul M. *Operational Record of the 95th Bomb Group (H)*

Barnes, Marvin E. *A History of the 452nd Bomb Group (H).* Privately Published

Bendiner, Elmer *Fall of the Fortresses.* Putnam, 1980

Birdsall, Steve *The B-17 Flying Fortress.* Morgan, 1965

Birdsall, Steve *Pride of Seattle: The Story of the first 300 B-17Fs.* Squadron Signal, 1998

Birdsall, Steve *Fighting Colors: B-17 Flying Fortress.* Squadron Signal, 1986

Birdsall, Steve *Hell's Angels: B-17 Combat Markings.* Grenadier Books, 1969

Birdsall, Steve & Freeman, Roger A. *Claims to Fame: The B-17 Flying Fortress.* Arms & Armour, 1994

Blakebrough, Ken *The Fireball Outfit: The 457th Bomb Group.* Aero, 1968

Bowden, Ray *Plane Names & Fancy Noses: The 91st Bomb Group (H).* Design Oracle Partnership, 1993

Bowers, Peter M. *Boeing Aircraft Since 1916.* Putnam, 1966

Bowers Peter M. *Fortress in The Sky.* Sentry, 1976

Bowman, M.W. *Four Miles High.* PSL, 1992

Bowman, M.W. *Flying To Glory.* PSL, 1992

Bowman, M.W. *Castles In The Air.* PSL, 1984

Bowman, M.W. *Boeing B-17 Flying Fortress.* Crowood, 1998

Bowman, M.W. with Tom Cushing *Confounding the Reich: The operational history of 100 Group RAF.* PSL, 1996

Bowman, M.W. & Woodall, Truett Lee Jr. *Helton's Hellcats: A Pictorial History of the 493rd Bomb Group.* Turner, 1998

Bowman, M.W. *B-17 Flying Fortress: Units of the 8th AF (Part 1 & 2).* Osprey, 2000 and 2002

Byers, Roland *Flak Dodger.* Pawpaw, 1985

Caidin, Martin *Flying Forts: the B-17 in WWII.* Ballantine, 1968

Collison, Thomas *Flying Fortress: The Story of the Boeing Bomber.* Scribner, 1943

Davis, Larry *B-17 In Action.* Squadron/Signal, 1984

Doherty, Robert E. & Ward, Geoffrey D. *Snetterton Falcons: The 96th Bomb Group in WWII.* Taylor Publishing, 1989

Ethell, Jeffrey L. & Simonsen, Clarence *The History of Aircraft Nose Art WWI to Today.* Motorbooks, 1991

Freeman, Roger A. *B-17 Fortress At War.* Ian Allan Ltd 1977

Freeman, Roger A. with David Osborne *The B-17 Flying Fortress Story.* Arms & Armour, 1998.

Freeman, Roger A. *The Mighty Eighth.* MacDonald, 1970

Freeman, Roger A. *The Mighty Eighth In Art.* Arms & Armour, 1996

Freeman, Roger A. *Airfields of the Eighth – Then and Now.* After the Battle, 1978

Freeman, Roger A. *Mighty Eighth War Manual.* Jane's, 1984

Good Brown, James *The Mighty Men of the 351st – Heroes All* Publishers Press, 1984

Havelaar, Marion H. with Hess, William N. *The Ragged Irregulars of Bassingbourn.* Schiffer Military History, 1995

Hawkins, Ian L. *B-17s Over Berlin: Personal Stories from the 95th Bomb Group (H).* Brasseys, 1990

Hess, William *B-17 Flying Fortress.* Ballentine, 1974

Huntzinger, Edward J. *The 388th At War.* 1979

Jablonski, Edward *Flying Fortress.* Doubleday, 1965

Johnsen, Fred *Winged Majesty.* PNAHF, 1980

Kurtz, Margo *My Rival The Sky.* Putnam, 1945

Le Strange, Richard, assisted by Brown, James R. *Century Bombers: The Story of the Bloody Hundredth.* 100th Bomb Group Memorial Museum, Thorpe Abbotts 1989

Lloyd, Alwyn T. and Moore, Terry D. *B-17 Flying Fortress in detail scale.* Aero publishers Inc

Members of the 490th Bomb Group (H) *Historical Record of the 490th Bomb Group (H).* Unpublished

McDowell, Ernest R. *Flying Fortress in Action.* Squadron Signal, 1987

MacKay, Ron. *351st Bomb Group.* Squadron Signal, 1994

McDowell, Ernest R. *Flying Fortress: The Boeing B-17.* Squadron Signal publications, 1987

McLachlan, Ian *Final Flights.* PSL, 1989

McLachlan, Ian *Eighth Air Force Bomber Stories.* PSL, 1991

Rust, Kenn C. *Eighth Air Force Story, WWII.* Historical Aviation Album, 1978

Rust, Kenn C. *Fifteenth Air Force Story, WWII.* Historical Aviation Album, 1976

Scutts, Jerry *B-17 Flying Fortress.* PSL, 1982

Siefring, T.A. *US Airforce in WWII.* Chartwell, 1977

Slater, Harry E. *Lingering Contrails of The Big Square A: A History of the 94th Bomb Group (H) 1942–45.* 1980

Smith Jnr, Ben *Chick's Crew: A Tale of the Eighth Air Force.* Privately published, 1978

Smith, John N. *Airfield Focus 37: Deenthorpe; 40: Molesworth; 44: Grafton Underwood.* GMS, 1999–2001

Stapfer, Hans-Heiri *Strangers in A Strange Land.* Squadron Signal, 1988

Streetley, Martin *Confound and Destroy: 100 Group and the Bomber Support Campaign.* Macdonald and Janes, 1978

Strong, Russell A. *First Over Germany: A History of the 306th BG.* Hunter Publishing, 1982

34th Bomb Group (H) Turner Publishing, 1988

Thompson, Charles D. *The Boeing B-17E and the F Flying Fortress.* Profile, 1966

The Story of the 390th Bomb Group (H). Privately printed, 1947

Varnedoe, W.W. Jr. *B-17s of the 385th Bomb Group, 8th AF.* Privately published, 1996

Varnedoe W.W. Jr. *A Navigator in World War II.* Privately published, 1997

Wilmott, H.P. *B-17 Flying Fortress.* Chartwell, 1980

Index

Page numbers in *italics* refer to illustrations.